Errata in: "Environmental factors and cultural measures affecting
the nitrate content in spinach" by T. Breimer.

(Page numbers without square brackets refer to journal; page numbers within
square brackets refer to book.)

Page 219 [29],   the figure presented here is Figure 3.2, see page 220 [30]
for legend.

Page 220 [30],   the figure presented here is Figure 3.1, see page 219 [29]
for legend.

Page 245 [55],   heading of Table 11: read "on" for "in".

Page 251 [61],   equation in section 5.4.1: read "(0–60 cm)" for "(6–60
cm)", "(0–60 cm)" for " (0–6 cm)"; and "harvested" for
"havested".

Environmental factors and cultural measures affecting the nitrate content in spinach

ENVIRONMENTAL FACTORS AND CULTURAL MEASURES AFFECTING THE NITRATE CONTENT IN SPINACH

T. BREIMER

1982

MARTINUS NIJHOFF/DR W. JUNK PUBLISHERS

Distributors

for the United States and Canada

Kluwer Boston, Inc.
190 Old Derby Street
Hingham, MA 02043
USA

for all other countries

Kluwer Academic Publishers Group
Distribution Center
P.O.Box 322
3300 AH Dordrecht
The Netherlands

Library of Congress Cataloging in Publication Data

Breimer, T.
    Environmental factors and cultural measures affecting
the nitrate content in spinach.

    Bibliography: p.
    1. Spinach.   2. Spinach--Composition.   3. Nitrates.
I. Title.
SB351.S7B73 1982   635'.41   82-10580

ISBN-13: 978-90-247-3053-7   e-ISBN-13: 978-94-009-7687-0
DOI: 10.1007/978-94-009-7687-0

Reprinted from *Fertilizer Research,* vol. 3, no. 3 (1982)

Preface

The present investigation was carried out in the period 1977-1981 at the Department of Soil Science and Plant Nutrition of the Agricultural University Wageningen, Netherlands. This university supplied the major financial support through a research grant, with additional financial support rendered by the Dutch Nitrogen Fertilizer Industry and Unilever Research, Netherlands. This support is gratefully acknowledged. In the present form this report is the author's doctoral dissertation submitted to the Faculty of the Agricultural University.

The author hereby wishes to thank prof.dr.ir. A. van Diest for his critical reading and correction of the English text, and dr.ir. J.H.G. Slangen for his valuable advices during the investigations and in the course of the preparation of this report. The author also acknowledges the valuable support received from several Agricultural Research Institutes and University Departments which made facilities and manpower available to conduct the experiments and to transform the results obtained into the present report.

Contents

Abstract

Environmental factors and cultural measures affecting the NO_3-content in spinach were studied indoors, in water-, sand- and soil-culture experiments. In the field, the influences of variations in N-fertilizing practices and in spinach varieties were also tested.

High NO_3-contents in spinach were found with low light intensities, with low soil-moisture contents, and with high temperatures. NO_3-contents increased with increasing K-dressing (less so with KCl than with K_2SO_4), but decreased with increasing soil pH. In pot experiments, positive results were obtained with sulphur-coated urea, with farmyard manure and with pig-manure slurry.

Application of Mo as a spray onto spinach leaves, and variations in P-dressings and in soil P-status were found not to affect the NO_3-content in spinach.

In pot experiments, NO_3-contents decreased with progressing plant age (in autumn less so than in spring). Within spinach plants, NO_3-contents were highest in petioles and older leaves. Varietal differences in NO_3-contents were observed in a pot- and a field experiment.

In pot- and field experiments, partial or complete replacement of NO_3-N by NH_4-N in general caused the NO_3-content in spinach to decrease. However, such a replacement was shown not always to result in lower NO_3-contents. Additional factors involved are e.g. the use and effectiveness of nitrification inhibitors, the soil type and the amount of available N.

The amount of N added and, in the field, the amount of N available in the soil before sowing, strongly affected the NO_3-content in spinach. Under field conditions, nitrogen appeared to be taken up from the top 60 cm of the soil profile.

The effects of variations in timing of nitrogen applications were absent in a pot experiment and not consistent in field experiments.

List of abbreviations

A = sum of inorganic anions ($Cl + H_2PO_4 + SO_4 + NO_3$) in meq per kg DM ($H_2PO_4$ stands for total P)

ADI = acceptable daily intake

C = sum of inorganic cations ($Ca + Mg + Na + K$) in meq per kg DM

C-A = excess of inorganic cations over inorganic anions in plants, meq per kg DM; NH_4-ions are not included

CEC = cation exchange capacity

DCD = dicyandiamide ($C_2H_4N_4$)

DM = dry matter

FW = fresh weight

FYM = farmyard manure

JECFA = Joint FAO/WHO Expert Committee on Food Additives

N-serve = nitrification inhibitor, containing 2-chloro-6-(trichloromethyl) pyridine ($C_6H_3NCl_4$), known as nitrapyrin, as active compound

org N = organic nitrogen (= total N - NO_3-N)

PMS = pig-manure slurry

SCU = sulphur-coated urea

WHC = water-holding capacity

WHO = World Health Organization

In the text ionic species are represented by their chemical symbols, charge signs being omitted, e.g. NO_3 instead of NO_3^-.

1 Introduction

Nitrate is present in many foodstuffs either as a natural component or as an additive. The main sources of dietary nitrate are vegetables and drinking water [Loggers, 1979; Tremp, 1981; Walker, 1975; White, 1975]. In the Netherlands, the mean daily intake of nitrate, estimated from 'total diet' studies, is about 160 mg NO_3 per 11000 kJ, 75% of which comes from vegetables and less than 10% from beverages and drinking water. There is a considerable variation in daily intake depending on amount and composition of the vegetables ingested [Loggers, 1979].

Although the toxicity of nitrate is relatively low, its occurrence in food must be considered hazardous because of the possible reduction of nitrate to nitrite that can occur before ingestion, in the gastro-intestinal tract or in saliva. Direct ingestion of nitrite can originate from additives of nitrate or nitrite to meat, fish and cheese or from bacterial or enzymatic reduction of nitrate in e.g. vegetables due to improper handling or storage [Tannenbaum, 1979]. In the Netherlands the estimated mean direct daily intake of nitrite is about 6 mg NO_2 [Loggers, 1979].

After ingestion of a given dose of dietary nitrate this passes to and through the stomach to be absorbed from the small intestine. Most of the absorbed nitrate is rapidly excreted in urine, but a part of it is secreted in the saliva, where in the presence of bacteria it is partly reduced to nitrite [Harada et al., 1975; Ishiwata et al., 1975; Spiegelhalder et al., 1976; Stephany & Schuller, 1978]. The extent of nitrite formation in saliva is related to the quantity of nitrate, to the concentration of the nitrate source and to the oral microflora [Tannenbaum et al., 1976]. On an average about 6% of the nitrate ingested is reduced to nitrite, which means that per 160 mg nitrate (NO_3) about 10 mg will be reduced in saliva to produce about 7 mg nitrite (NO_2) [Stephany & Schuller, 1980]. Another possible environment for nitrate reduction is the human stomach. Nitrite concentrations in a normal human stomach are generally lower than 1 mg per litre [Tannenbaum, 1979]. There are, however, conditions under which the human stomach can acquire nitrite concentrations of tens or even hundreds of mg per litre [Ruddell et al., 1976; Tannenbaum et al., 1979]. This occurs when the pH of the stomach, which is normally quite low, rises to a level high enough to allow the growth of bacteria which can reduce nitrate to nitrite. Under these conditions, sometimes occurring in infants under three months of age and in patients with specific gastric disorders, the stomach pH becomes similar to that of the mouth and the stomach may even contain bacteria similar to those found in the mouth. It was recently suggested that also under normal conditions nitrate and nitrite may be formed endogenously in the human intestine, probably by small bowel bacterial flora [Tannenbaum et al., 1978].

Nitrite is much more toxic than nitrate. Acute nitrite poisoning causes methemoglobinemia; infants under three months of age are very susceptible to this disease [Kübler & Simon, 1969]. Cases of infant methemoglobinemia in Germany have been reported to originate mainly from drinking water high in nitrate and from improperly stored baby food containing nitrite [Simon, 1970]. Methemoglobin can completely revert to hemoglobin through methemoglobin-reductase and symptoms of methemoglobinemia will disappear within a few hours.

More concern should be given to the fact that nitrite might cause chronic poisoning as a result of reacting in vivo with secondary nitrogen compounds occurring naturally in certain foods, to form N-nitroso-compounds. When administered to laboratory animals, these compounds are potent carcinogens [Magee & Barnes, 1967; Preussmann, 1981], but as yet they have not been definitely incriminated as being the cause of any human cancer [Fraser et al., 1980].

N-nitroso-compounds can be formed endogenously in humans; they have been identified in vivo in the stomach, in the infected urinary bladder and in saliva and faeces [Fraser et al., 1980]. In epidemiological studies, gastric cancer risk was linked to nitrogen (fertilizer) usage in Chile [Armijo & Coulson, 1975; Zaldivar, 1977; Zaldivar & Wetterstrand, 1975] and to the nitrate concentration in well-water in Colombia [Cuello et al., 1976; Tannenbaum et al., 1979]. Case-control studies in Japan [Haenszel et al., 1976] and among Japanese migrants to Hawaii [Haenszel et al., 1972] have linked gastric cancer risk to certain food items, and to well-water use in Japan [Haenszel et al., 1976]. In the latter two studies, however, a decreased gastric cancer risk was associated with food items such as raw, green, leafy vegetables and fresh fruit, rich in vitamin C, a known inhibitor of N-nitrosation [Mirvish, 1977].

An epidemiological study in Iran has considered the possibility of a link between dietary nitrate and cancer of the oesophagus [Joint Iran-International Agency for Research on Cancer Study Group, 1977]. Other studies on cancer mortality in relation to water supplies as the source of ingested nitrate have been carried out in the United Kingdom [Hill et al., 1973] and in Illinois, USA. [Geleperin et al., 1976]. Fraser et al. [1980] in a review of the above-mentioned studies state that, based on the present epidemiological evidence, the hypothesis that high nitrate ingestion is involved in the aetiology of gastric cancer, should not be lightly discarded. To their opinion, however, there is too little information available to draw any conclusions on the relationship between high nitrate ingestion and any other form of human cancer.

In view of the possible risk of high nitrate and nitrite intakes the European Office of the World Health Organisation (WHO) has recommended a maximum nitrate concentration in drinking water of 50 mg per litre [WHO,

1970] and the Joint FAO/WHO Expert Committee on Food Additives (JECFA) set maximum acceptable daily intakes (ADI) of nitrate and nitrite. For adults of 60 kg body weight the ADI for nitrate is 220 mg [JECFA, 1974] and the ADI for nitrite is 8 mg [JECFA, 1976]. Although the latter standards are set only for additives and do not account for the 'natural' amounts of nitrate and nitrite in foods and for the conversion of nitrate to nitrite in saliva, it seems reasonable to denote them as the total acceptable daily intakes [Stephany & Schuller, 1978].

In the Netherlands the mean (total) daily intake of nitrite (13 mg NO_2) exceeds the ADI set by the JECFA, and the ADI for nitrate is sometimes exceeded as well [Loggers, 1979]. Since the high nitrate intake is mainly of vegetable origin, the amounts of nitrate in vegetables should be reduced [Aldershoff, 1982; Mol, 1979]. Furthermore, because of the possible risk of methemoglobinemia, vegetables used as baby food should be extremely low in nitrate.

A review of nitrate contents in a wide range of vegetables [Corré & Breimer, 1979] revealed that vegetables such as table beet, celery, chervil, lamb's lettuce, lettuce, purslane, radish, spinach and turnip tops frequently have high nitrate contents. From these vegetables spinach, which is also frequently used as baby food, was chosen as a test crop.

The aim of the present work was to study environmental factors affecting the nitrate content in spinach and to find cultural measures to reduce these contents. Spinach grown for the processing industry was studied mainly.

2 Literature

2.1 Nitrogen transformations in soil

Most of the nitrogen in soil is incorporated in organic matter, with only a very small proportion of soil nitrogen being directly available to plants. This available N occurs mainly in the form of nitrate and ammonium. The major processes in soil which generate plant-available nitrogen can be identified as biological nitrogen fixation and ammonification, the latter being a process in the mineralisation of organic matter.

In arable soils of the temperate regions, non-symbiotic nitrogen fixation is generally low [Mengel & Kirkby, 1978] and the extent of symbiotic nitrogen fixation depends on the inclusion of leguminous crops in the crop rotation system. In mineral soils of the temperate regions, generally less than 3% of the total organic nitrogen is ammonified annually [Allison, 1973]. Soil type and organic matter composition are primary factors influencing the quantity of ammonium released [Barker & Mills, 1980] and environmental conditions such as low temperatures and deficiency or excess of water can limit ammonification.

In most arable soils, the ammonium formed as a result of organic matter decomposition is rapidly converted into nitrate in a stepwise process called nitrification which is also biologically controlled. The rate at which nitrification takes place is determined by environmental factors such as soil water and soil oxygen content, pH and temperature [Huber & Watson, 1974]. The amounts of available nitrogen normally released from mineral soils are inadequate to meet the nitrogen requirements of most cultivated crops.

In horticulture, it is generally recognised that nitrogen fertilizer is needed to produce yields or growth rates considered to be economical under most soil- and cropping conditions [Barker & Mills, 1980]. When soil- and climatic conditions are favourable, fertilizer nitrogen applied in other than the nitrate form is rapidly converted to nitrate. Only with the use of 'controlled' or 'slow release' fertilizers and with nitrogen carriers such as manures and sewage sludge will the conversion rate be less rapid. When, however, environmental conditions are unfavourable, the nitrification process can be strongly suppressed, leading to an accumulation of ammonium whose rate of formation from soil organic matter or from a fertilizer, like urea, is affected much less by adverse environmental conditions. Also with the aid of so-called nitrification inhibitors, nitrification can be suppressed [Gasser, 1970; Kerkhoff & Slangen, 1980; Prasad et al., 1971].

2.2 Nitrate uptake

Both nitrate and ammonium can be taken up and assimilated by plants, but since nitrate is the prevailing form of available nitrogen in soil, most nitrogen is taken up as nitrate.

The mechanism by which nitrate is absorbed by plant roots appears to be a very complex one in that it is influenced by a number of environmental factors affecting both the external supply of nitrate and the physiological and biochemical processes operating within the plant [Barker & Mills, 1980]. Both nitrate uptake and -reduction in higher plants are reported to be substrate-induced processes [Beevers & Hageman, 1969; Jackson, 1978]. As nitrate has been shown to induce nitrate reductase, a relationship between nitrate reductase and uptake is likely to exist [Huffaker & Rains, 1978; Jackson, 1978; Schrader, 1978]. Nitrate uptake is known to increase sharply with increases in the external supply of nitrate. According to Huffaker & Rains [1978], more than one nitrate uptake system may operate in plants. The uptake of nitrate appears to be influenced little by other anions such as chloride, bromide or sulphates [Rao & Rains, 1976], but cations such as calcium, potassium and ammonium affect nitrate uptake significantly [Jackson, 1978; Rao & Rains, 1976]. Increases in the supply of calcium and potassium generally accelerate the rate of nitrate uptake, whereas ammonium ions have an inhibitory effect [Haynes & Goh, 1978].

Nitrate absorption is apparently sensitive to the external hydroxyl-ion concentration. Rao & Rains [1976] observed a decline in nitrate uptake above pH 6, but no decline at pH values as low as 4.0. The decline in nitrate uptake at high pH values was explained by inhibition of the transport system by hydroxyl. For the maintenance of nitrate uptake a continual supply of energy appears to be essential. Carbon dioxide is known to exert an inhibitory effect on nitrate uptake. According to Barker & Mills [1980] this may be due either to the competition between the processes of carbon dioxide reduction and nitrate uptake for energy, or to stomatal closure in the presence of carbon dioxide. Nitrate absorption is suppressed by low and promoted by high temperatures [Frota & Tucker, 1972; Lycklama, 1963].

2.3 Nitrate reduction in plants

Although there is evidence that most plants reduce nitrate in both roots and leaves, species differ in the location at which the majority of the nitrate is reduced. In most plant species, however, more nitrate reduction occurs in the leaves than in the roots [Beevers & Hageman, 1969]. Olday et al. [1976] observed that about 70% of the total nitrogen in the xylem of spinach was nitrate-nitrogen which implies that most nitrate is transported to the leaves.

Nitrate reduction appears to be a very complex biochemical process which is affected by a number of factors [Beevers & Hageman, 1969; Hewitt, 1975]. Evidence exists that in most plant tissues a compartimentalisation of nitrate takes place [Jackson, 1978]. The vacuole is assumed to be the storage pool of nitrate [Aslam et al., 1976] and the metabolic or active pool of nitrate is thought to be in the cytoplasm where reduction of nitrate to nitrite is catalysed by the enzyme nitrate reductase. Nitrate reductase, a molybdo-flavo-protein, is thought to be the rate limiting enzyme in the pathway for reduction of nitrate to ammonium [Beevers & Hageman, 1969]. The enzyme nitrate reductase is adaptive and inducible by nitrate in cytoplasm. In various tissues, induction is related to the capacity for protein synthesis and thus to maturation and age of the tissue [Beevers & Hageman, 1969].

Light plays an important role in nitrate reduction. In their review, Beevers & Hageman [1972] conclude that light effects are exerted at two stages, one consisting of the provision of reductant for nitrate to nitrite reduction, and the other of the control of the level of the enzyme required for nitrate reduction. Diurnal fluctuations of nitrate reductase activity have been shown in various plants [Schrader, 1978; Srivastava, 1980].

Nitrate reduction and photosynthesis seem to be closely related [Beevers & Hageman, 1972]. The influence of carbon dioxide on nitrate assimilation, however, is not completely understood [Schrader, 1978].

Nitrate reductase activity was shown to be also temperature dependent. Inactivation at temperatures lower than the optimum one is not as drastic as at temperatures higher than the optimum one [Pflüger & Wiedemann, 1977]. The regulation of nitrate reductase activity in higher plants has been recently reviewed by Srivastava [1980]. Evidence presented in this review suggests that enzyme activity may be decreased by e.g. the presence of ammonium, by water stress and by molybdenum-, calcium- and sulphur deficiencies. It was reported that there is a possibility that nitrate reductase activity is genetically regulated [Beevers & Hageman, 1969; Srivastava, 1980].

2.4 Nitrate contents in plants

The quantity of nitrate in plants (or in plant parts) results from the difference in nitrate uptake (and transport) on the one hand, and nitrate reduction on the other hand. Hence, the factors regulating these processes also affect the quantity of nitrate present in a plant.

On account of the fact that the nitrate content is the ratio between the quantity of nitrate and the quantity of dry matter, it follows that the nitrate content in a plant also depends on factors which influence the production of dry matter.

Literature on nitrate in vegetables is extensive. Recent reviews have been published by Corré & Breimer [1979], Maynard et al. [1976] and Venter [1978].

In the following, the literature on cultural measures and environmental factors affecting the nitrate contents of mainly spinach will be reviewed.

[8]

2.5 Cultural measures

2.5.1 Nitrogen dressing

2.5.1.1 Nitrogen amount

The influence of nitrogen dressings on the nitrate contents in spinach has been studied in several countries in indoor experiments conducted both in greenhouses and growth chambers [Barker, 1975; Barker & Maynard, 1971; Barker et al., 1974; Cantliffe, 1972-1; Cantliffe, 1972-2; Cantliffe, 1972-3; Cantliffe, 1973-1; Cantliffe, 1973-2; Cantliffe & Phatak, 1974-2; Cantliffe et al., 1974; Dressel & Jung, 1970; Dressel, 1976; Grujic & Kastori, 1974; Hildebrandt, 1976; Hulewicz & Mokrzecka, 1971; Jacquin & Papadopoulos, 1977; Jurkowska, 1971; Jurkowska & Wojciechowicz, 1974; Kick & Massen, 1973; Maynard & Barker, 1971; Maynard & Barker, 1972; Maynard & Barker, 1974; Merkel, 1975; Mills et al., 1976; Olday et al., 1976; Roorda van Eysinga & van der Meys, 1980-1; Schudel et al., 1979; Siegel & Vogt, 1974; Siegel & Vogt, 1975-1; Siegel & Vogt, 1975-2; Terman et al., 1976; Terman & Allen, 1978] and outdoor experiments [Acar & Ahrens, 1978; Achtzehn & Hawat, 1969; Aworh et al., 1980; Barker et al., 1971; Becker, 1965; Bengtsson, 1968; Boek & Schuphan, 1959; Brink et al., 1968; Brown & Smith, 1967; Brown et al., 1969; van Burg et al., 1967; van Burg et al., 1969; Descamps, 1972; Dressel & Jung, 1970; Eerola et al., 1974; Hansen, 1978; Knauer, 1970; Knauer & Simon, 1968; Kuhlen, 1962; Lambeth et al., 1969; Lee et al., 1971; Lorenz & Weir, 1974; van Maercke, 1973; Maga et al., 1976; Maynard & Barker, 1979; Mehwald, 1973; Nicolaisen & Zimmermann, 1968; Schuphan, 1965; Schuphan, 1974; Schuphan & Hentschel, 1970; Schuphan et al., 1967; Schütt, 1977; Tronicková & Vit, 1970; Tronicková & Vit, 1972; Witte, 1970; Zimmermann, 1966].

In both indoor- and outdoor experiments nitrate contents were found to increase with increasing N-dressings, which indicates that the nitrate contents in spinach are strongly influenced by the nitrate concentration in the growth medium. In outdoor experiments, the nitrate contents generally increased with N-dressings up to 200 kg per ha, but declined at higher dressings. Spinach yields, in general, showed optima which mostly were attained at N-dressings at which nitrate contents had not yet started to decline.

Large differences in nitrate content and yield of plants resulting from a certain N-dressing were found between experiments as well as within one experiment. One reason for the different results obtained among field experiments might be that the amount of nitrogen which is available in soil before the experiment started is not accounted for. The importance of knowledge on residual available nitrogen in soil was demonstrated by Böhmer [1980]. In field experiments, she found that both residual available nitrogen and applied fertilizer nitrogen strongly affected the yields of spinach and other vege-

[9]

tables, which can be considered as a first indication that for a complete evaluation of the nitrogen nutrition of vegetables it is important to have a knowledge of the quantity of available nitrogen in the soil. Nitrate contents in the crop, however, were not determined in this study. Other factors which may explain differences observed in response to fertilizer nitrogen are the form in which the nitrogen is applied, the time of application, the age of the plants, the quantity and quality of soil organic matter, soil pH, rainfall, temperature and other environmental conditions.

2.5.1.2 Nitrogen form

It has already been postulated that in most arable soils under normal conditions ammonium will be rapidly nitrified. Hence, it is to be expected that urea, ammonium and nitrate fertilizers will give similar results when applied before sowing of a spinach crop [Barker et al., 1971; Dressel & Jung, 1970; Knauer & Simon, 1968; van Maercke, 1973; Siegel & Vogt, 1974; Tronicková & Vit, 1970; Tronicková & Vit, 1972]. Differences between these N-fertilizer forms, however, can occur when N-dressings are high or when environmental conditions are such that nitrification is retarded. During the relatively short period of growth of a spinach crop, nitrate fertilizers will effectuate under such circumstances higher nitrate concentrations in soil resulting in higher nitrate contents in plants. This might also explain why in pot experiments, where the soil volume is relatively small, nitrate contents in spinach are often found to be lower with ammonium than with nitrate as applied N-form [Jung & Dressel, 1978; Kick & Massen, 1973; Mills et al., 1976].

With controlled or slow-release nitrogen carriers and organic nitrogen carriers, the release of ammonium is the rate-limiting step and therefore nitrate concentration in soil in general will be lower with these N-carriers than with other N-sources for equal quantities of N applied. The significance of controlled-release fertilizers for horticultural crops has been recently reviewed by Maynard & Lorenz [1979]. In an indoor experiment, Siegel & Vogt [1975-1] demonstrated that two controlled-release fertilizers produced significantly lower nitrate contents in spinach than did calcium nitrate or urea. In a field experiment, Schuphan et al. [1967] found that with the use of a controlled-release fertilizer (Peraform), nitrate contents in spinach were reduced by more than 50% compared with equal dressings of N in the form of calcium nitrate. Yield of spinach, however, was reduced by about 20%.

Organic nitrogen and inorganic nitrogen carriers were compared by Barker [1975], Jacquin & Papadopoulos [1977], Maga et al. [1976], Schudel et al. [1979], Schuphan [1974] and Siegel & Vogt [1975-2]. In general, at equal N-dressings, nitrate contents and spinach yields were lower with organic nitrogen forms. Obviously, nitrogen mineralisation (ammonification) was slow enough to result in lower nitrate concentrations in soil than encountered with the use of inorganic nitrogen forms. The effect of differences in mineralisa-

[10]

tion among organic nitrogen carriers was shown in an indoor experiment by Barker [1975]. With dried cow manure which mineralised much more slowly than four other organic nitrogen carriers, yields and nitrate contents of spinach were significantly depressed.

Next to environmental conditions, nitrification inhibitors can also suppress nitrification. Well-known nitrification inhibitors are N-serve (active compound: 2-chloro-6-(trichloromethyl)pyridine ($C_6H_3NCl_4$), known as nitrapyrin) and dicyandiamide ($C_2H_4N_4$, known as DCD). Nitrification inhibitors added to ammonium sulphate or urea in indoor experiments resulted in decreases in nitrate contents of spinach [Jung & Dressel, 1978; Jurkowska, 1971; Jurkowska & Wojciechowicz, 1974; Kick & Massen, 1973; Mills et al., 1976; Moore, 1973; Siegel & Vogt, 1975-2; Sommer & Mertz, 1974], but yield in general also decreased. The effect of nitrification inhibitors on the nitrate content and yield of spinach in a field experiment was demonstrated by Bengtsson [1968]. He found that addition of nitrapyrin to ammonium sulphate (2% by weight of N-dose) reduced the nitrate contents by about 50% and yields by about 15%. Moore et al. [1977] studied the effect of ammonium sulphate with and without nitrapyrin in field, greenhouse and growth chamber experiments. They stated that nitrapyrin addition increased yield and depressed the nitrate contents. With the use of ammoniacal nitrogen sources nitrification inhibitor additions tend to decrease both nitrate contents and yields. In a growth chamber experiment with spinach, Mills et al. [1976] therefore investigated whether a partial replacement of ammonium by nitrate in combination with increasing nitrapyrin doses could result in both high yields and low nitrate contents. From these results they concluded that a fertilizer mixture of 50% ammonium-nitrogen and 50% nitrate-nitrogen, especially in conjunction with the use of nitrapyrin, was the best combination for maximum growth and minimum nitrate contents. Recent literature on nitrification inhibitors and their effects was reviewed by Kerkhoff & Slangen [1980].

2.5.1.3 Time of application of nitrogen

In the Netherlands, in commercial spinach growing, all nitrogen is usually applied before sowing. Only after periods of low temperatures and/or excessive rain, nitrogen is side-dressed [Buishand, 1974].

The effects of variations in time of application of nitrogen were studied in field experiments with spinach by Barker et al. [1971], Kuhlen [1962], Maga et al. [1976], Mehwald [1973], Nicolaisen & Zimmermann [1968] and Zimmermann [1966]. In general, it was observed that yield decreased and nitrate contents increased with side-dressed N in comparison with equal amounts of N applied in basal dressings. Comparing basal- and side dressing of ammonium nitrate, only Barker et al. [1971] found that basal dressing resulted in higher nitrate contents. Furthermore, in a comparison of side dressings of different forms of N-fertilizer applied nine days before harvest,

[11]

these investigators observed that at final harvest potassium nitrate had produced higher nitrate contents than had ammonium nitrate and urea. In their experiments, however, all yields were alike.

2.5.2 Other nutrients

Nitrate contents in spinach were found to be only slightly affected by phosphate dressings, whereas yield in general increased [Brown et al., 1969; Cantliffe, 1973-1; Regan et al., 1968; Schuphan, 1965; Terman et al., 1976; Terman & Allen, 1978]. In most experiments, potassium was observed to increase the nitrate contents and yields of spinach [Boek & Schuphan, 1959; Brown et al., 1969; Cantliffe, 1973-1; Nurzynski, 1976; Prummel, 1971; Regan et al., 1968]. In some experiments, however, high potassium dressings suppressed nitrate contents [Grujic & Kastori, 1974; Knauer & Simon, 1968].

In a sand culture the abstinence of either phosphate, potassium, calcium or magnesium 21 days before harvest was studied by Barker & Maynard [1971]. Yield and nitrate content of spinach were not significantly affected by these treatments. According to Maynard et al. [1976], there is no evidence that sodium, calcium and magnesium have any direct effect on the nitrate contents in vegetables. In a field experiment carried out by Brown et al. [1969] and Regan et al. [1968], liming suppressed the nitrate content and increased the yield of spinach. The effect of liming on the nitrate contents was a function of the size of the phosphate dressing. As a result of liming soil pH-H_2O rose from 5 to 6. An increase in both nitrate content and yield with increase in pH, however, were observed in an indoor experiment by Cantliffe et al. [1974]. In indoor experiments molybdenum applications to molybdenum-deficient soils were found to decrease the nitrate contents in spinach [Cantliffe et al., 1974; Hildebrandt, 1976]. Deficiencies of manganese and copper increased the nitrate contents and decreased yield [Hildebrandt, 1976]. Also boron deficiency seems to increase the nitrate contents in spinach [Hulewicz & Mokrzecka, 1971].

Although it has been postulated that nitrate uptake is little influenced by variations in the supply of other anions, there is evidence that lower nitrate contents in spinach will be obtained with potassium chloride than with potassium sulphate [Boek & Schuphan, 1959; Nurzynski, 1976; Schmalfusz & Reinicke, 1960].

2.5.3 Herbicides

The effects of herbicides on weed control, yield and nitrate contents of spinach were studied by Cantliffe & Phatak [1974-1]. Three herbicides, cycloate, alachlor and lenacil gave satisfactory weed control and yields, but the nitrate contents in spinach of plots treated with these three herbicides

[12]

were 3 or 4 times higher than in spinach of hand-weeded or non-weeded control plots. The other herbicides under investigation did not give satisfactory weed control, but in general raised the nitrate contents in spinach also. According to the authors the herbicides, especially cycloate, alachlor and lenacil, are increasing NO_3-accumulation through a decrease in some phase of nitrate reduction. Singh et al. [1972] observed that herbicides with s-triazines like simazin and triazin, as active components, did not affect the nitrate contents of bush beans and spinach.

2.5.4 Spinach variety

Differences in nitrate contents among spinach cultivars have been observed by Barker [1975], Barker et al. [1971], Barker et al. [1974], Cantliffe [1972-2], Cantliffe [1972-3], Cantliffe [1973-2], Cantliffe & Phatak [1974-2], Maynard & Barker [1972], Maynard & Barker [1974] and Olday et al. [1976]. In these studies, nitrate contents in savoy-leaved cultivars were found to be higher than in smooth-leaved types. Semi-savoyed types were generally intermediate between these extremes, but individual cultivars sometimes overlapped into one or the other group. According to Olday et al. [1976], the difference in nitrate contents between smooth- and savoy-leaved cultivars must be ascribed to differences in nitrate assimilation rather than to variations in rates of uptake or transport. They observed that nitrate reductase activity in a smooth-leaved cultivar was 2 to 3 times higher than in a savoy-leaved cultivar. Terman & Allen [1978], re-examining the results of Barker et al. [1974], however, suggested that the differences in nitrate contents found by the latter authors might have to be attributed to differences in dry-matter production of the cultivars.

Differences in nitrate contents between spinach cultivars were furthermore studied by Cools et al. [1980], Descamps [1972], van Maercke & Vereecke [1976], Pavlek et al. [1974], Schuphan et al. [1967], Tronicková & Vit [1970], Tronicková & Vit [1972] and Westvlaamse Proeftuin [1978]. In most experiments, differences among cultivars were noticed, but the results obtained were not always consistent.

2.5.5 Harvesting procedure

Nitrate contents vary among plant parts of spinach, high contents being found in petioles and lower contents in leaf blades [Achtzehn & Hawat, 1971; Barker et al., 1971; Barker et al., 1974; Bodiphala & Ormrod, 1971; Cools et al., 1980; Descamps, 1972; Dressel & Jung, 1970; Eerola et al., 1974; Lorenz & Weir, 1974; Olday et al., 1976]. Expressed on a dry-weight basis, the nitrate contents in petioles and leaf blades may differ by a factor of 10. Schuphan et al. [1967] reported that with harvesting methods yielding chiefly leaf blades, the nitrate contents in the harvested spinach would be about 60%

lower. However, they did not indicate how much yield would be reduced.

Within one plant, nitrate contents are generally higher in older tissue [Barker & Maynard, 1971; Bodiphala & Ormrod, 1971] and in general it can be stated that, if other circumstances remain unchanged, the nitrate content in spinach will increase with age. However, during the cropping season soil nitrate will decline which, depending on N-dressing, can be expected to result in a decrease in nitrate content with age. Weather conditions (light, temperature, rain) also exert their effects so that the nitrate content in spinach can vary considerably during a growth period [Descamps, 1972]. Nitrate contents in spinach increasing as well as decreasing with time have been found in outdoor experiments [Aworh et al., 1980; van Maercke, 1973; Nicolaisen & Zimmermann, 1968; Schuphan et al., 1967; Witte, 1970; Zimmermann, 1966]. In indoor experiments with pots where the soil volume is limited and weather conditions play a less important role, a decrease in nitrate content in spinach with age is generally to be expected.

2.6 Environmental conditions

2.6.1 Light

Light plays an important role in both nitrate assimilation (section 2.3) and plant growth. It is one of the main factors determining the nitrate contents in plants. Light intensity, photoperiod and possibly also the time expired within the photoperiod affect the nitrate content in a plant. In field studies it was observed that nitrate contents in shaded spinach plants were significantly higher than those in unshaded plants [Klett, 1968; Schuphan et al., 1967], and in growth chamber studies high nitrate contents were associated with low light intensities [Boek & Schuphan, 1959; Cantliffe, 1972-1; Cantliffe, 1973-1; Dressel & Jung, 1970; Tychsen, 1976]. Cantliffe [1972-2] demonstrated that the nitrate content in beet root decreased with increasing length of the photoperiod. In the same study, it was observed that the nitrate contents of 2 spinach cultivars grown with a high nitrogen supply decreased about 50% during a 12-hour photoperiod. A third cultivar, however, did not show this effect. With a low nitrogen supply, no changes during the photoperiod were found with any of the cultivars. A decrease in nitrate content during the photoperiod was also observed by Tychsen [1976] in young spinach plants and by Minotti & Stankey [1973] in young beet root plants. Schwerdtfeger [1975] reported that samples of field-grown spinach and head lettuce harvested in the afternoon contained less nitrate than did samples harvested in the morning. When light intensity was reduced by shading no differences were found. No diurnal variation in nitrate content of greenhouse-grown head lettuce, harvested in March, was observed by Roorda van Eysinga & van der Meys [1980-2].

2.6.2 Carbon dioxide

Carbon dioxide plays a role in both nitrate uptake and nitrate assimilation. According to Maynard & Barker [1979] there is evidence that high nitrate contents may occur in plants growing in atmospheres low in carbon dioxide.

2.6.3 Temperature

Literature concerning temperature effects on the nitrate contents in plants is ambiguous. This may be due to the inability to distinguish between temperature effects on nitrate uptake and on nitrate reduction or between effects of temperature on soil nitrogen dynamics and on plant nitrogen metabolism [Maynard & Barker, 1979]. In a growth chamber experiment, Cantliffe [1972-3] noticed that the nitrate contents in spinach increased with increasing temperatures up to $25^{o}C$ and decreased between 25 and $30^{o}C$. A significant interaction between temperature and nitrogen supply was found with respect to the nitrate contents. In this experiment, temperature will have affected both mineralisation and nitrification.

2.6.4 Water

Water stress is known to produce increased nitrate contents in plants [Wright & Davison, 1964]. The increase in accumulation of nitrate is explained as resulting from a decrease in nitrate reductase activity prior to the moment at which uptake starts to decline. Although low soil moisture availability will be an exception in vegetable growing, Maynard et al. [1976] suggested that short dry periods might lead to higher nitrate contents. They further postulated that nitrate contents might increase with increasing atmospheric humidity due to reduced transpiration rates resulting in a decreased translocation of nitrates to the induction sites. Evidence to confirm these postulates is lacking, however. Also through its effects on ammonification and nitrification soil moisture will exert an influence on the nitrate supply to plants.

2.6.5 Season

Seasonal variations in nitrate contents are connected with temperature and especially with daylength and light intensity. In spring, daylength and light intensity increase during the development of a crop, whereas in autumn they decrease. Soil temperature, however, is relatively low in spring and relatively high in autumn, while air temperature on the average is the same. In general, for dry-matter production and nitrate reduction spring conditions are more favourable than are autumn conditions. As a result, with comparable N-dressings therefore nitrate contents are generally lower in spring

[Achtzehn & Hawat, 1969; Eerola et al., 1974; Knauer & Simon, 1968; van Maercke, 1973; Mehwald, 1973; Schuphan et al., 1967; Schütt, 1977; Witte, 1967; Witte, 1970].

2.6.6 Location

Vegetables grown at different locations may show different nitrate contents [Wedler, 1979]. These differences due to location can be a reflection of difference in soil type or in the climatic conditions varying among regions in which a crop is grown. Since processes such as mineralisation, nitrification, nitrogen leaching and denitrification are largely determined by the type of soil, the nitrate contents in vegetables will be affected by the soil on which they are grown [Geyer, 1978]. The nitrate contents in vegetables grown on one soil type but in different regions may vary due to regional differences in daylength, irradiation, temperature, rainfall etc. [Boek & Schuphan, 1959; Knauer & Simon, 1968]. Differences in irradiation and temperature also exist between a greenhouse- and an outdoor situation.

3 Materials and methods

3.1 Indoor experiments

3.1.1 General

In the indoor experiments, the effects of variations in N-supply, $NO_3:NH_4$-ratio, time of application of N, type of N-fertilizer, nitrification inhibitors, plant age, light intensity, temperature, soil-moisture content, soil pH, Mo-, P- and K-supply and variety on yield and chemical composition of spinach were studied.

The experiments can be subdivided in: water-culture, sand-culture and soil-culture experiments. Most of these were carried out in glasshouses, one in growth chambers.

In the experiments with quartz sand or soils, pots (with a content of 6.9 l) and techniques described by Schuffelen et al. [1952] were employed.

3.1.2 Water-culture experiments

The uptake of major plant nutrients by spinach plants supplied with nitrate-N or without nitrogen, was studied with a continuous-flow (water-culture) system.

Spinach seedlings (cv. Subito) were grown on quartz sand moistened with a dilute nutrient solution. When two leaves had emerged (18 December 1979), seedlings were transplanted to 6-l containers (10 plants per container) with the flowing nutrient solution. The composition (in mmol per l) of the nutrient solution was: Na 1, K 5, Ca 1, Mg 1.5, H_2PO_4 1, NO_3 5, Cl 2 and SO_4 1.5. Furthermore the solution contained trace elements, including Fe added as Fe-EDTA. The pH of the solution was about 5.5. The flow-rate per container was about 0.2 l per hour. After periods of 36, 45, 50, 56, 62 and 65 days following transplanting, sets of 30-40 plants were harvested (3 or 4 containers per harvest). Shoot- and root material were separated.

On 12 and 18 February 1980 (56 and 62 days after transplanting), plants grown on the NO_3-solution were transferred to a solution without NO_3 (NO_3 was replaced by SO_4). Plants of these treatments were harvested on 15, 18 and 21 February 1980 (3 containers per harvest).

The experiment was carried out in a slightly heated glasshouse with an average temperature of 15°C.

[17]

3.1.3 Sand-culture experiments

In pot trials with quartz sand as growth medium, the effects of variations in N-supply and in NO_3:NH_4-ratio on yield and chemical composition of spinach were studied in 1978 (A) and 1980 (B).

Basal dressings (in g per pot) were: Na 0.15, P 0.44, Mg 0.24, $CaCO_3$ 10. Sufficient amounts of trace elements and Fe-EDTA were supplied and the nutrients were mixed with 5 kg sand, which was covered with 1.5 kg unfertilized sand as a germination layer. Per pot, 40 seeds of spinach were sown. After emergence, the number of plants per pot was reduced to 20.

Trial site: unheated glasshouse.

A. Variety: Estivato. Sowing date: 21 April. Emergence date: 27 April. Harvest dates: 27, 34, 37 and 39 days after emergence.
   K-dressing per pot: 1.08 g K.
   N-dressing per pot: 1.5 g N (0.75 g on 20 April and 0.75 g on 14 May).
   NO_3-N:NH_4-N-ratios: 3:0, 2:1, 1:2 and 0:3.
   N-serve 24E per pot: 110 mg (= 26 mg nitrapyrin).

B. Variety: Subito. Sowing date: 4 September. Emergence date: 7 September. Harvest dates: 37, 40, 43 and 46 days after emergence.
   K-dressings per pot: (for the NO_3-N:NH_4-N-ratio 3:0 only) 1.08 and (for all ratios) 1.83 g K (1.08 g on 4 September and 0.75 g on 29 September).
   N-dressings per pot: 1.5 and 2.25 g N (0.75 g on 4 and 0.75 g on 25 September for both N-levels and 0.75 g on 6 October for high N).
   NO_3-N:NH_4-N-ratios: 3:0, 2:1 and 1:2.
   N-serve 24E per pot: 100 mg (= 24 mg nitrapyrin).

In both trials, NH_4-N was applied as $(NH_4)_2SO_4$ and NO_3-N as $Ca(NO_3)_2$. Per harvest, the aerial parts and roots of 3 pots were collected separately per pot.

3.1.4 Soil-culture experiments

In the pot experiments with different soils as growth medium the effects of several factors were studied.

Experiment 1, 1977

Effects of variations in plant age and N-supply on the distribution of plant nutrients in the aerial parts of spinach.

Growth medium: a sandy soil (Wageningen-A, Table 1), limed to a pH-KCl of about 5.7. Variety: Estivato. Sowing date: 1 August. Emergence date: 7 August.

Harvest dates: 24, 32, 38, 45, 52, 59 and 66 days after emergence.

N-dressings per pot: 0.5 and 2.25 g NO_3-N.

Table 1. *Characteristics of the soils used in the soil-culture experiments (3.1.4) and in the field experiments (3.2)*

	Duiven A	Duiven B	Duiven C	Lelystad A	Lelystad B	Nagele	Helden A	Helden B	Wageningen A	Wageningen B	Droevendaal	Vredepeel
<2 μm (%)	35	30	36	22	23	–	4	–	–	–	–	–
<16 μm (%)	–	–	65	–	32	–	5	–	–	–	–	–
org. C (%) a)	1.5	1.8	1.7	0.9	1.4	1.1	1.4	1.7	1.6	1.6	1.5	2.2
$CaCO_3$ (%)	1.3	0.1	0.6	5.5	7.6	–	–	–	–	–	–	–
pH-KCl	6.8	5.9	6.7	7.3	7.4	7.3	5.7	5.9	5.0	3.8	5.4	4.8
N_{tot} (g per kg)	1.7	2.2	2.0	–	1.0	–	1.0	1.0	–	–	–	–
P_w b)	38	41	39	8	9	162	45	59	43	–	22	–
P-AL c)	170	157	–	–	–	116	284	350	389	219	280	210
K-HCl d)	174	199	208	100	174	–	133	174	116	100	158	133
Mg-NaCl e)	17	31	34	–	53	–	72	72	36	–	–	42
CEC f)	18.6	–	23.0	–	11.8	–	4.2	–	–	–	–	–

a) Oxidation with dichromate according to Kurmies.
b) P extractable with water according to Sissingh [1971] and expressed as P in mg per l soil.
c) P extractable with ammarium lactate-acetic acid and expressed as P in mg per kg soil.
d) K extractable with 0.1 M HCl and expressed as K in mg per kg soil.
e) Mg extractable with 1 M NaCl and expressed as Mg in mg per kg soil.
f) Determined with neutral ammonium acetate and expressed in meq per 100 g soil.

Trial site: covered glasshouse. Temperature (day/night): 15°C. Light intensity (400-700 nm): 50 W per m^2 (HPLR). Daylength: 12 hours. Relative humidity: 70-90%.
Leaves were sampled in pairs, according to age, and divided into laminae and petioles.

Experiment 2, 1979

Effects of variations in plant age and N-supply on yield and chemical composition of spinach.

Growth medium: a sandy soil (Helden-B, Table 1). Variety: Spartan. Sowing date: 23 August. Emergence date: 29 August. Trial site: unheated glasshouse.

Harvest dates: 13, 19, 27, 33, 40 and 47 days after emergence.

N-dressings per pot: 0 and 1.5 g NO_3-N.

Experiment 3, 1979 (A) and 1980 (B)

Effects of variations in plant age, N-supply and time of application of N on yield and chemical composition of spinach.

Growth medium: a sandy soil (Helden-B, Table 1). Variety: Spartan. Trial site: unheated glasshouse.

N-dressings per pot: 0 (N_0) - 0.5 (N_1) - 0.75 (N_2) - 1.0 (N_3) - 1.0 + 0.25 (N_{3+1}) - 1.0 + 0.5 (N_{3+2}) - 1.25 (N_4) - 1.25 + 0.25 (N_{4+1}) -1.5 (N_5) - 1.75 (N_6) and 2.0 (N_7) g NO_3-N. The basal dressings were applied before sowing and for the higher amounts the second dressing was applied about two weeks after emergence. The additional N-dressings of 0.25 g per pot, ($N_{3+1,3+2}$) and ($N_{3+2,4+1}$) were applied at the first and third harvest, respectively.

A. Sowing date: 4 September. Emergence date: 11 September.

Harvest dates: 29 ($N_{3,4,5}$), 35 ($N_{3,3+1}$), 42 ($N_{3,3+1,4,5}$), 49 ($N_{3,3+1,3+2,4,4+1,5}$) and 56 (all treatments) days after emergence.

B. Sowing date: 28 February. Emergence date: 17 March.

Harvest dates: 37 ($N_{3,4,5}$), 42 ($N_{3,3+1}$), 45 (all treatments, except $N_{3+2,4+1}$), 50 ($N_{3,3+1,3+2,4,4+1,5}$) and 53 ($N_{3,3+1,3+2,4,4+1,5}$) days after emergence.

Experiment 4, 1978 (A) and 1979 (B)

Effects of variations in N-supply and soil-moisture contents on yield and chemical composition of spinach.

A. Growth medium: a clay loam (Duiven-A, Table 1). Variety: Spinoza. Sowing date: 8 March. Emergence date: 15 March. Harvest date: 27 April. Trial site: unheated glasshouse.

N-dressings per pot: 0 - 0.4 - 0.8 - 1.6 - 3.2 g NO_3-N.

B. Growth medium: a loam (Nagele, Table 1). Variety: Estivato. Sowing date: 7 March. Emergence date: 22 March. Harvest date: 14 May. Trial site: unheated glasshouse.

N-dressings per pot: 0 - 0.4 - 0.8 - 1.2 - 1.6 g NO_3-N.

In both trials soil-moisture contents were: 40, 60 (standard for all other pot trials) and 80% of the maximum water-holding capacity (WHC).

For most Dutch soils, 60% of the maximum water-holding capacity is nearly pF 2.0.

Table 2. *Chemical composition of the organic soil amendments in soil-culture experiments 4 and 8.*

Material	Constituents, in % of dry weight (A and B) or g per liter (C)						
	N	P	K	Ca	Mg	NO_3-N[b]	NH_4-N[b]
A. Sewage sludge	2.0	1.1	0.2	1.8	0.2	–	–
B. Farmyard manure	2.3	1.3	1.9	2.9	0.6	0.01	–
C. Pig manure slurry[a]	4.8	1.1	4.1	1.8	0.6	0	3.5

[a] Volumic weight is 1.0 kg per liter and dry matter content is 4.9%.
[b] Extraction with 0.5 M acetic acid.

Experiment 5, 1978 (A) and 1979 (B)

Effects of variations in N-supply and type of N-carriers on yield and chemical composition of spinach.

N-dressings per pot: 0 - 0.8 - 1.6 - 2.4 g N for calcium nitrate (15.5% N) and sulphur-coated urea (SCU: 37% N), and 0 - 2.4 - 4.8 - 7.2 g N for urea-form (-aldehyde) (= Peraform: 37% N) and sewage sludge (see Table 2). It was assumed that during the growth period of the crop only one third of the nitrogen from ureaform and sewage sludge would become available.

A. Growth medium: a clay loam (Duiven-A, Table 1). Variety: Spinoza. Sowing date: 8 March. Emergence date: 15 March. Harvest date: 27 April. Trial site: unheated glasshouse.

B. Growth medium: a loam (Nagele, Table 1). Variety: Estivato. Sowing date: 7 March. Emergence date: 22 March. Harvest date: 14 May. Trial site: unheated glasshouse. In trial B per pot 60 mg N-serve 24E was added (= 14 mg nitrapyrin).

Experiment 6, 1980

Effects of variations in N-supply, type of fertilizer and nitrification inhibitors on yield and chemical composition of spinach.

Growth medium: a sandy soil (Droevendaal, Table 1). Variety: Spartan. Sowing date: 1 September. Emergence date: 7 September. Harvest date: 24 October. Trial site: unheated glasshouse.

[21]

N-dressings per pot: 0 - 1.0 - 1.5 - 2.0 g N.
Fertilizers: $Ca(NO_3)_2$, $(NH_4)_2SO_4$ and $CaCN_2$ (18% N, 1.5% NO_3-N). For incubation, the fertilizers were mixed with the soil on 20 August.
In addition to all treatment combinations without inhibitor, $(NH_4)_2SO_4$ (1.0 - 1.5 - 2.0 g N) was treated with 30 and 90 mg dicyandiamide (DCD) or nitrapyrin (N-serve) and $Ca(NO_3)_2$ (1.0 - 1.5 -2.0 g N) with 30 mg DCD.

Experiment 7, 1978

Effects of variations in N-supply, NO_3:NH_4-ratio, light intensity and temperature on yield and chemical composition of spinach.

Growth medium: a clay loam (Duiven-A, Table 1). Variety: Estivato. Sowing date: 13 January. Emergence date: 18 January. Trial site: growth chambers. Temperature until 15 February (day/night): 17°C. Light intensity until 15 February (400-700 nm): 62 W per m^2. Day length: 12 hours. Relative humidity: 80%.
N-dressings per pot: 1.5 and 2.25 g N.
NO_3-N:NH_4-N-ratios: 10:0, 7:3 and 4:6.
N-serve 24E per pot: 130 mg (= 31 mg nitrapyrin). First harvest date: 14 February. On 15 February, 9 temperature and light intensity combinations were installed in the growth chambers.
Temperatures: 22, 17 and 12°C.
Light intensities: 70, 49 and 33 W per m^2.
Second harvest date: 24 February.

Experiment 8, 1980

Effects of variations in N-supply and of inorganic vs. organic nitrogen on yield and chemical composition of spinach.

Growth medium: a sandy soil (Vredepeel, Table 1). Variety: Spartan. Trial site: unheated glasshouse.
On 13 December 1979, portions of the soil were mixed with either pig-manure slurry (PMS) (3 or 6 litre per 100 kg dry soil), or dried farmyard manure (FYM) (1.25 or 2.5 kg per 100 kg dry soil) (see Table 2). One portion remained without manure. The samples without manure and with pig-manure slurry were limed on 10 January and on 20 February 1980, respectively. During incubation, soil-moisture content was kept at about 50% of the maximum water-holding capacity.
N-dressings per pot for all samples (mixtures) were: 0 - 0.5 - 1.0 - 1.5 g NO_3-N; 2.0 g NO_3-N was included only for the treatment without manure. Sowing date: 10 March. Emergence date: 24 March. Harvest date: 12 May. Plants of the treatments 0 - 1.0 - 2.0 g NO_3-N without manure, and 0 - 1.0 g NO_3-N with 6 liter PMS per 100 kg dry soil were also harvested on 22 April and on 1, 6 and 15 May.

Experiment 9, 1980

Effects of variations in N-supply, soil pH and type of fertilizer on yield and chemical composition of spinach.

Growth medium: a sandy soil (Wageningen-B, Table 1). Variety: Spartan. Sowing date: 5 March. Emergence date: 19 March. Harvest date: 6 May. Trial site: unheated glasshouse.

N-dressings per pot: 0.4 - 0.8 - 1.2 - 1.6 - 2.0 g N.

Soil pH: 4.5 - 5.0 - 5.5 - 6.0 (pH-KCl).

Fertilizers: $Ca(NO_3)_2$ for all N-levels and $(NH_4)_2SO_4$ for 1.6 g N only. To establish the desired soil pH-values, $CaCO_3$ was mixed with the soil. $CaSO_4$ was added to keep the amounts of Ca constant.

Experiment 10, 1978

Effects of variations in N-supply, soil type and of Mo-spraying on yield and chemical composition of spinach.

Growth medium: a sandy soil and a clay loam (Wageningen-A and Duiven-A, Table 1). The sandy soil was limed to a pH-KCl of about 5.7. Variety: Estivato. Sowing date: 4 August. Emergence date: 10 August. Harvest date: 21 September. Trial site: unheated glasshouse.

N-dressings per pot: 0 - 0.75 - 1.5 g NO_3-N.

Mo-treatments: no spraying, spraying on 15 September or spraying on 19 September. Plants were sprayed with a solution containing 100 mg Mo per l until all leaves were wet.

Experiment 11, 1979

Effects of variation in P-supply on yield and chemical composition of spinach.

Growth medium: a clay loam (Duiven-B, Table 1). Variety: Spinoza. Trial site: unheated glasshouse.

On 29 November 1978, portions of the soil were mixed with increasing amounts of $CaHPO_4$. During incubation, soil-moisture contents were kept at about 50% of the maximum water-holding capacity. P-AL contents of the soil analysed in March 1979 were: 157, 214, 284 and 380 mg P per kg soil. Each of these soil samples was dressed per pot with 0 - 0.22 - 0.44 - 0.66 g P as $CaHPO_4$ on 2 March.

Sowing date: 2 March. Emergence date: 12 March. Harvest date: 1 May.

N-dressing per pot: 1.6 g N as NH_4NO_3.

Experiment 12, 1979 (A) and (B) and 1980 (C)

Effects of variations in K-supply and type of K-carriers on yield and chemical composition of spinach.

Growth medium: a sandy soil (Wageningen-A, Table 1) with K-HCl-contents of 35, 75, and 69 mg K per kg soil.

K-dressings per pot: 0 - 0.42 - 0.83 - 1.66 g K.

K-carriers: KCl and K_2SO_4.

N-dressings per pot: 1.2 (C) and 1.6 (A and B) g N as $Ca(NO_3)_2$.

A. Variety: Spartan. Sowing date: 10 August. Emergence date: 15 August. Harvest date: 13 September. Trial site: unheated glasshouse.

B. Variety: Spartan. Sowing date: 24 September. Emergence date: 1 October. Harvest date: 20 November. Trial site: heated glasshouse (15°C).
C. Variety: Subito. Sowing date: 8 January. Emergence date: 15 January. Harvest date: 19 March. Trial site: heated glasshouse (15°C).

Experiment 13, 1980

Effects of variation in spinach variety on yield and chemical composition of spinach.

Varieties: Nobel, Virkade, Medania, Califlay.
Sowing date: 21 February. Planting date: 25 February. Harvest dates: 17 and 24 April. Trial site: heated glasshouse (17°C).
In this experiment plastic 1.5-l pots were used with about 1 kg of a commercial potting mixture. Based on its wet weight, the potting mixture contained 71% water, 21% organic matter, 8% ash and per kg of substrate 0.38 g (NO_3+NH_4)-N (KCl-extraction). pH-H_2O was 6.0 and available P, K and Mg were high.
Per pot 1 plant was grown. The experiment was arranged in 4 blocks with 5 pots per block.

The experiment was carried out at the Institute for Horticultural Plant Breeding, Wageningen. Varieties were kindly supplied by dr.ir. A.H. Eenink of this Institute.

Remarks: In the experiments with N-, P-, or K-supplies as variables the other nutrients were applied in sufficient amounts before sowing; high amounts of N and K were applied in split dressings. In all experiments 40 seeds of spinach were sown per pot and after emergence the number of plants per pot was reduced to 20. There were at least 3 replicates (= sometimes blocks) in all experiments. Residual (NO_3+NH_4)-N in the pots was determined in some of the experiments.

3.2 Field experiments

3.2.1 General

In all field experiments, the effects of variation in NO_3-N dressings on yield and chemical composition of spinach were studied. Available (NO_3+NH_4)-N was determined at the start and throughout the growth period, on fallow, untreated plots. To prevent the growth of spinach and other plants these fallow plots were regularly hoed. In most of the experiments, residual soil N was measured immediately after the harvest.

In addition to the effects of varying N-dressings, in a part of the field experiments the effects of variations in $NO_3:NH_4$-ratio and of the addition of nitrapyrin (active component of N-serve) on yield and chemical composition of spinach were studied. In two trials, the effects of other factors, e.g. time of application of N and varieties were tested.

To realize the various $NO_3:NH_4$-ratios, calcium nitrate and ammonium sulphate were used as single fertilizers or in combination with each other. Nitrapyrin, dissolved in an ample volume of acetone, was sprayed on the fertilizer grains, and left to dry. The amounts of nitrapyrin used were based on the amounts of nitrogen applied. The nitrogen (with or without nitrapyrin) was broadcast and, except for the Duiven experiments (section 3.2.2), worked in mechanically immediately thereafter. The amount of seed was in all trials 50 kg per ha and to prevent weed growth, after sowing (and before emergence) 6-7 l Asulox per ha was sprayed.

3.2.2 Duiven experiments

These experiments on a river clay loam soil (Duiven A, B and C, Table 1) were carried out in the spring periods of 1977, 1978 and 1979, at the Agricultural Experiment Station of Unilever Research, Duiven.

Experiment 1 (A and B), 1977
This experiment was laid out as a randomized block with three replicates; plot size 5x5 m^2.
On 5 April P (83 kg P as superphosphate) and K (415 kg K as K-60) were applied per ha. Sowing date: 20 April. Variety: Estivato. Emergence date: 2 May.
Sprinkler irrigation on 11 May (7.5 mm), 26 May (7.5 mm) and 1 June (9 mm).
Harvest dates: 2 June, 8 June and 15 June (morning and evening). The first harvest was planned one week before stalk elongation, the second at the beginning of stalk elongation and the third at 5 cm stalk length. In fact, the second and third harvests were one or two days late, at stalk length of about 2-4 cm and 7-10 cm, respectively.
A. N-dressing per ha: 150 kg N.
   NO_3-N:NH_4-N-ratios: 0:3, 1:2, 2:1 and 3:0.
   Nitrapyrin: none and 45 ml N-serve 24E per kg N (= about 1% nitrapyrin).
   Time of application of N: all N applied immediately after sowing (D_1) on 26 April, 3 split applications (D_3) on 26 April, 10 and 24 May and 6 split applications (D_6) as D_3 and on 3, 12 and 31 May.
B. N-dressings per ha: 0 - 75 - 150 (as in A) - 225 kg NO_3-N, applied on 26 April.
   Soil sampling dates: 26 April, 2, 16, 23 and 31 May and 6 and 13 June, on fallow plots without nitrogen dressings.

Experiment 2 (A and B), 1978
In this experiment A and B were laid out separately as randomized blocks with four replicates; plot size 5x7 m^2.
On 5 April P (83 kg P as superphosphate) and K (320 kg K as potassium sulphate) were applied per ha.
Sowing date: 14 April. Variety: Estivato. Emergence date: 28 April. Harvest date: 2 June (morning and evening).

[25]

A. N-dressing per ha: 100 kg NO_3-N, applied on 13 April.

Additional N-dressings per ha: 0 - 50 - 100 kg NO_3-N, 50-100 kg NH_4-N or 50 + 50 kg NO_3-N + NH_4-N. These additional dressings were treated with 89 ml N-serve 24E per kg N (= about 2% nitrapyrin) and applied on 11 May.

B. N-dressings per ha: 0 - 50 - 100 - 150 - 200 kg NO_3-N, applied on 17 April.

Soil sampling dates: 5 April, 2, 16 and 29 May and 2 June on fallow plots without nitrogen dressings, and after harvest on all plots.

Experiment 3, 1979

This trial was laid out as a split plot scheme, replicated four times with the N-levels randomized per block and the varieties randomized per N-level; plot size 1.5 x 10.5 m^2.

On 7 May, P (83 kg P as superphosphate) and K (332 kg K as K-40) were applied per ha.

Sowing date: 9 May. Emergence date: 20 May. Harvest dates: 13 June (variety 1), 15 June (variety 2), 19 June (varieties 3, 4 and 5).

N-dressings per ha: 0 - 50 - 100 - 150 - 200 kg NO_3-N, applied on 9 May. Varieties: Indian Summer (1), Summic (2), Medania (3), Symfonie (4), Nr. 731037 (5).

Soil sampling dates: 10 and 29 May, 5, 13 and 20 June on fallow plots without nitrogen dressings.

3.2.3 Lelystad experiments

These experiments were carried out in the autumn of 1978 and in the spring of 1979 on a loamy soil (Table 1) at the Research Station for Arable Farming and Field Production of Vegetables, Lelystad. In both years the trials were laid out as randomized blocks with four replicates; plot size 5x5 m^2.

Experiment 4, 1978

On 31 July K (50 kg K as K-60) was applied per ha. P was not applied as the preceding crop being seed potatoes had received an ample quantity of P-fertilizer.

Sowing date: 11 August. Variety: Spartan. Emergence date: 18 August. Harvest date: 26 September.

N-dressings per ha: 0 - 50 - 100 - 150 - 200 kg N, applied on 31 July (50, 100 and 200 kg per ha as NO_3-N only).

NO_3-N:NH_4-N-ratios (for the 150 kg N-dressing): 0:3, 1:2, 1:1, 2:1 and 3:0.

Nitrapyrin: 100 ml N-serve 24E per kg N (= about 2% nitrapyrin) for all ratios; without nitrapyrin for the ratios 0:3 and 3:0.

Soil sampling dates: 31 July, 21 August, 5, 19 and 26 September on fallow plots without nitrogen dressings. After the harvest, plots with NO_3-N only (without nitrapyrin) were also sampled.

Experiment 5, 1979
On 23 March, P (66 kg P as triple superphosphate) and K (66 kg K as K-60) were applied per ha.
Sowing date: 27 April. Variety: Estivato. Emergence date: 10 May. Harvest date: 18 June.
The experimental design was the same as in 1978 with an additional N-treatment of 150 kg per ha, NO_3-N : NH_4-N = 1:1, without nitrapyrin. N-application date: 19 April.
Nitrapyrin: 84 ml N-serve 24E per kg N.
Soil sampling dates: 19 April, 15, 24 and 30 May, 6, 11, 14 and 18 June on fallow plots without nitrogen dressings. After the harvest, plots with NO_3-N only (without nitrapyrin) were also sampled.

3.2.4 Helden experiments

These experiments were carried out in spring and autumn 1979 on a sandy soil (Helden-A and B, Table 1) at the Experimental Horticulture Garden, Helden. In both seasons the trials were laid out as randomized blocks with four replicates; plot size 4x4 m^2.

Experiment 6, 1979 (spring)
On 5 April P (22 kg P as superphosphate), K (83 kg K as K-40) and Mg (30 kg Mg as kieserite) were applied per ha.
Sowing date: 11 April. Variety: Estivato. Emergence date: 23 April. Harvest date: 7 June.
N-dressings per ha: 0 - 50 - 100 - 150 - 200 kg N, applied on 10 April (50 and 200 kg per ha as NO_3-N only).
NO_3-N:NH_4-N-ratios (for the 100 or 150 kg N-dressings): 0:4, 1:3, 2:2, 3:1 and 4:0.
Nitrapyrin: 84 ml N-serve 24E per kg N (= about 2% nitrapyrin) for all ratios. Additionally, without nitrapyrin for the ratios 0:4, 2:2 and 4:0.
Soil sampling dates: 10 and 25 April, 9, 22, 28 and 31 May, 5 and 7 June on fallow plots without nitrogen dressings. After the harvest, plots with NO_3-N only (without nitrapyrin) were also sampled.

Experiment 7, 1979 (autumn)
The experimental design and the basal dressings were the same as in the spring.
Basal dressings applied: 20 August. Sowing date: 23 August. Variety: Spartan. N-application date: 21 August. Emergence date: 31 August. Harvest date: 15 October.

[27]

Soil sampling dates: 21 August, 3, 11, 17 and 25 September, 1, 8 and 15 October on fallow plots without nitrogen dressings. After the harvest, plots without nitrogen and with 100 and 200 kg NO_3-N (without nitrapyrin) per ha were also sampled.

3.3 Weather conditions

Daily precipitations recorded at Wageningen for April-July 1977, March-September 1978 and March-October 1979 are shown in Figure 3.1.

The rainfall at Wageningen and Duiven (30 km east of Wageningen) was similar during the cropping period. At Duiven 104, 53 and 104 mm were recorded in 1977, 1978 and 1979, respectively.

Precipitation recorded in 1978 and 1979 at Lelystad (70 km north of Wageningen) was somewhat higher than in Wageningen. Between emergence and harvest of the crop, 77 and 134 mm were recorded at Lelystad in 1978 and 1979, respectively. Rainfall at Helden (75 km south of Wageningen) was not recorded; it is assumed to have been the same as at the Experimental Horticulture Station at Meterik, 15 km north of Helden. During the cropping period in the spring of 1979, the 87 mm recorded at Meterik were lower than in Wageningen; in the autumn of 1979, precipitation was recorded at Meterik until 16 September only. Global radiation data per day in MJ per m^2 recorded at Wageningen in 1977 (April-July), 1978 (March-September), 1979 (March-November) and 1980 (February-November) are presented in Figure 3.2. The radiation during the periods of the indoor and field experiments can be read from this figure. Radiation was lower during the field experiments 4 and 7 (autumn) than during the field experiments in spring. The data on maximum and minimum temperatures (Table 3), demonstrate that in the course of the field experiments in spring the average daily temperature increased by about 10°C, whereas during the autumn experiments it decreased by about 5°C.

3.4 Harvest

In the field experiments in the spring the spinach was harvested at a stalk length of about 5 cm, whereas in the autumn, when stalk length could not be used as a criterion, harvest took place about 6 weeks after emergence.

The plants were harvested between 9.00 and 11.00 a.m. On 15 June 1977 and 2 June 1978 the 'evening' harvests were between 6.00 and 8.00 p.m. Of at least 2 m^2 in the centre of a plot the plants were cut off below the stem and leaves with a special knife. As opposed to harvesting with the mowing machine or scythe, this practice gives less sampling variation. Because of irregular growth, in Lelystad 1978 (experiment 4) 2x2 m^2 and in Duiven 1979 (experiment 3) 1x3 m^2 plots were harvested. The samples were stored at 2-4°C and weighed. Subsamples of 0.5-1.0 kg were weighed and dried at

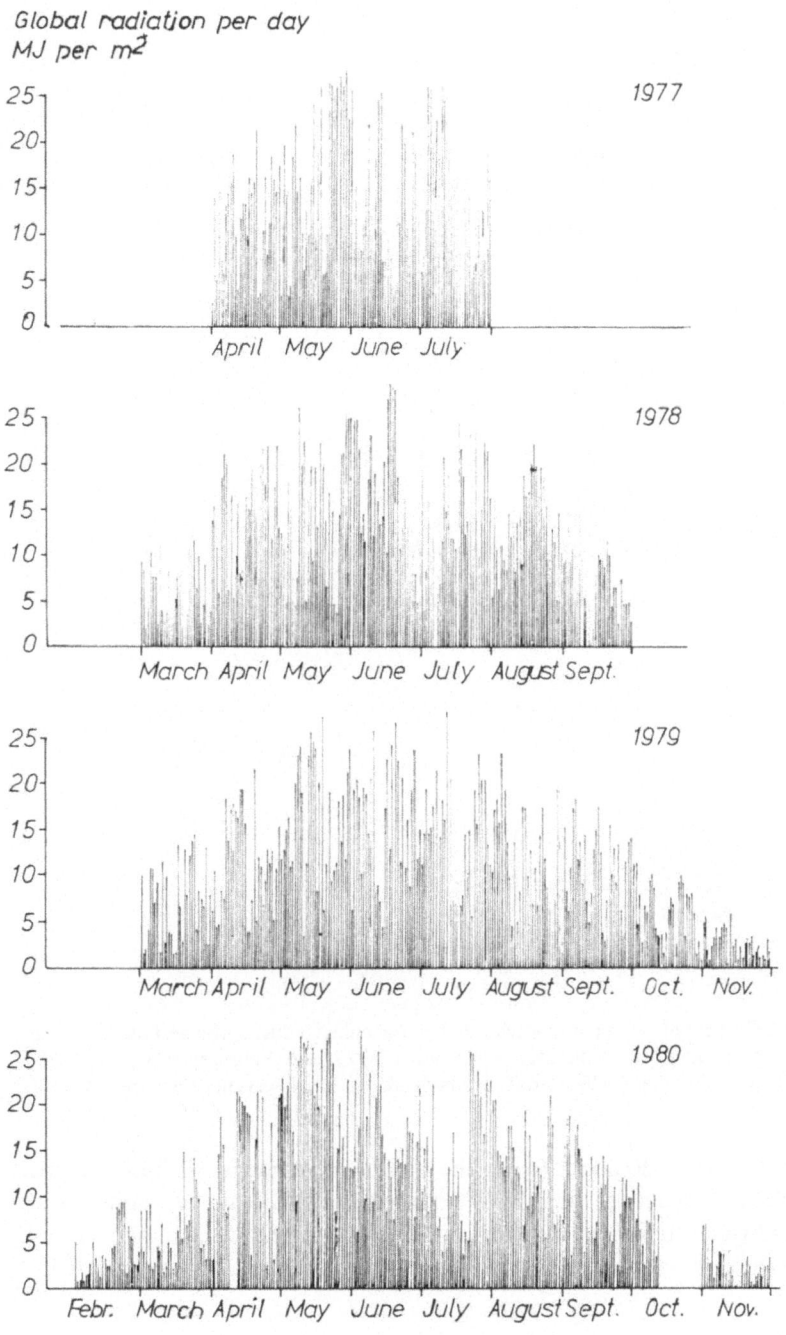

Figure 3.1 Precipitation per day recorded at Wageningen during the periods April-July 1977, March-September 1978 and March-October 1979.

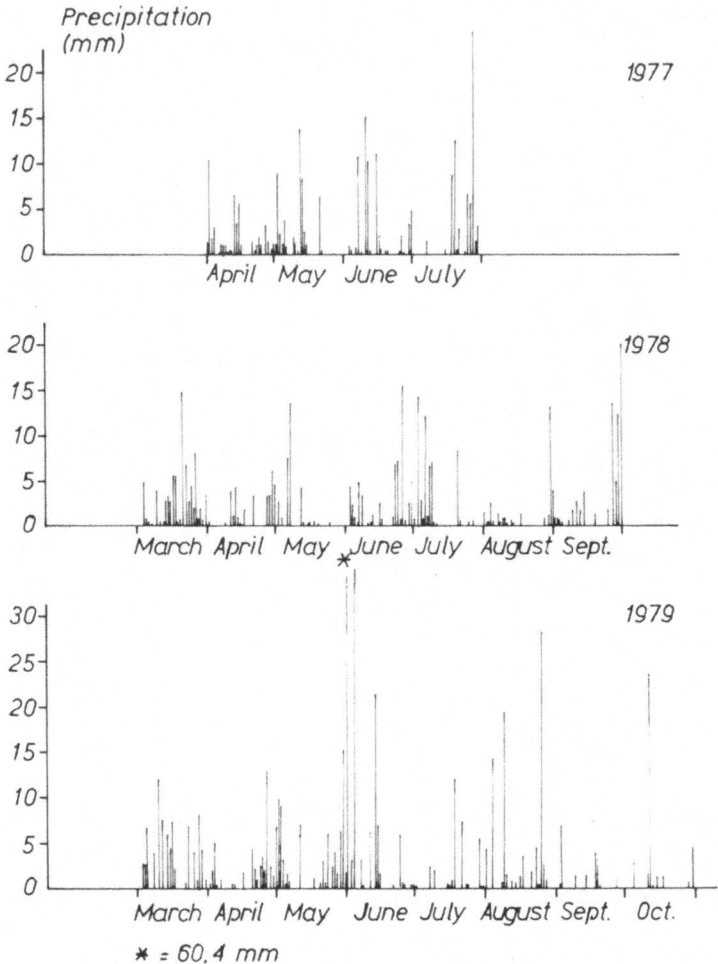

Figure 3.2 Global radiation per day recorded at Wageningen during the periods April-July 1977, March-September 1978, March-November 1979 and February-November 1980 (from 13 to 31 October 1980 no data available due to translocation of weather station).

70°C for at least 20 hours (from the indoor experiments all fresh material was dried). Subsamples of all 1979 experiments had to be washed with deionized water to remove soil particles. The dried plant material was weighed, ground through a 1-mm sieve and stored in air-tight plastic bags for chemical analysis.

3.5 Chemical analysis

Samples of the plant material were digested with concentrated sulphuric acid, salicylic acid and hydrogen peroxide. In the digests, Na, K and Ca were determined flame-photometrically, H_2PO_4 and total N colorimetrically and Mg by atomic absorption spectrophotometry.

Subsamples were extracted with water (0.5 g plant material and 50 ml deionized water) and in the extracts nitrate was determined with an 'Orion' ion-specific electrode, Cl coulometrically and SO_4 turbidimetrically. Details of the analytic procedures and techniques are given by Slangen & Hoogendijk [1970]. 0.5 M acetic acid was used instead of water for the samples of the field trials and in these extracts nitrate was determined colorimetrically as NO_2 after reduction with a Cd-Cu column in a Technicon Autoanalyzer (system II).

The analytical results were expressed as mmol per kg dried (70°C) plant material.

The carboxylate contents, (C-A) according to de Wit et al. [1963], were found by subtracting the sum of inorganic anions from the sum of inorganic cations both expressed as meq per kg. Organic N is total N minus NO_3, both in mmol per kg.

Table 3. *Mean maximum and minimum air temperatures per decade recorded at 1.5 m above groundlevel in Wageningen in 1977 (April-June) and in 1978 and 1979 (April-October)*

| Month | Decade | 1977 | | 1978 | | 1979 | |
		Max.	Min.	Max.	Min.	Max.	Min.
April	I	7.9	−1.2	11.0	1.0	9.9	1.6
	II	9.8	−0.7	9.9	−0.2	14.7	4.3
	III	13.2	5.6	15.7	4.0	10.8	3.5
May	I	13.6	5.5	16.8	6.7	11.0	1.6
	II	15.4	6.7	15.1	4.7	19.2	8.2
	III	19.7	6.0	19.7	8.7	19.1	9.8
June	I	17.5	8.0	23.1	12.5	21.4	11.7
	II	19.5	12.2	19.2	8.1	18.4	8.2
August	I			19.2	11.7	20.5	11.4
	II			21.4	9.6	19.8	11.5
	III			18.6	8.7	19.1	8.3
September	I			18.1	9.6	21.9	10.7
	II			17.0	8.2	17.4	7.9
	III			16.0	9.4	16.4	2.6
October	I			14.9	7.1	19.2	9.7
	II			15.0	6.0	15.0	7.5

3.6 Soil analysis

Soils on the field plots (or other units) were sampled in layers of 20 and 30 cm, to a depth of 80 and 90 cm, respectively. With a special auger 5-6 cores were taken per layer and the soil material of the cores was mixed thoroughly. To prevent changes in N-contents [Breimer & Slangen, 1981], while in the field, the samples were stored in a cool box and after transport to the laboratory they were kept at 2°C for a period not exceeding 2 days. Subsamples of approximately 20 g of the mixed field-moist soil were weighed (in duplicate) and shaken with 50 ml extraction solution, i.e. 0.01 M $CuSO_4$ for NO_3-N and 1 M KCl for NH_4-N. After 15 minutes ($CuSO_4$) or 60 minutes (KCl), the extracts were filtered (Schleicher and Schüll 589^3 Blue Ash Free). NO_3 was measured with an 'Orion' ion-specific electrode [Meyers & Paul, 1968; Øien & Selmer-Olsen, 1969] and NH_4 was determined colorimetrically [Novozamsky et al., 1974].

The moisture contents of the samples were determined simultaneously by drying about 20 g of soil at 105°C. NO_3-N and NH_4-N contents of the samples were expressed as mg per kg oven-dry soil (105°C). Bulk density was determined for each soil type and layer. The results ranged from 1.2 to 1.5 t per m^3. From these data, the inorganic N-contents (in kg per ha) could be calculated per layer and for the whole profile-depth.

4 Nitrogen nutrition and the ionic balance in spinach; results of the indoor experiments

4.1 General introduction

During the process of ion uptake by plants electroneutrality is maintained both in the plant and in the growth medium. As plant roots absorb nutrient ions at different rates, electroneutrality is achieved within the plant by accumulation or degradation of anions of organic acids [Hiatt, 1967; Jacobson & Ordin, 1954; Ulrich, 1941]. The difference between quantities of inorganic cations (C) and -anions (A) in the plant (C-A), henceforth referred to as carboxylate pool, is equivalent to the amount of these organic anions. In the nutrient medium, neutrality is maintained by excretion of either H- or HCO_3-ions from the roots [Kirkby, 1974].

Processes that generate carboxylates are cation uptake in excess of anion uptake, nitrate assimilation and sulphate assimilation [Dijkshoorn, 1969; Hiatt, 1967; Jacobson & Ordin, 1954; Ulrich, 1941, 1942]. Processes that deplete the carboxylate pool are anion uptake in excess of cation uptake, and ammonium assimilation [Yem & Folkes, 1954; Coic et al., 1961, 1962; Houba et al., 1971].

Plants supplied with nitrate usually excrete HCO_3- or OH-ions into the nutrient medium. Details of this process are outlined in the Ben Zioni-Lips model [Ben Zioni et al., 1971]. In grasses and cereals some 50-60% of the negative charge originating from NO_3-assimilation appears in the nutrient medium. In a number of plant species, e.g. sugar beet [Houba et al., 1971] and tomato [Kirkby & Knight, 1977] the OH-ion excretion is relatively small. Kirkby [1974] therefore proposed a modification of the Ben Zioni-Lips model that is applicable to species having a cation:anion uptake ratio close to unity. In this model, after translocation to the shoot, nitrate-nitrogen is transformed into organic nitrogen, and carboxylates accumulate in the shoot. Translocation of carboxylates to the roots and subsequent decarboxylation in the root system are thought to be of minor importance in these species.

Plants supplied with ammonium-nitrogen as only N-source absorb cations in excess of anions [Kirkby, 1969], and they assimilate ammonium mainly in the roots. The intracellular pH of these plants is regulated by H-ion excretion in the nutrient medium [Raven & Smith, 1976], thus lowering the pH of the rhizosphere. According to Kirkby [1969], the H-ion excretion by these plants is equivalent to (C-A) (including free NH_4-ions) + (organic N) equivalents - (organic S) equivalents. Carboxylate production of ammonium-fed plants generally is low [Breteler, 1973; Houba et al., 1971; Kirkby, 1969].

Van Egmond [1978] suggests that when plants are supplied with both ammonium- and nitrate-nitrogen, for certain plants, e.g. sugar beet, spinach and tomato, the extent of ammonium- or nitrate nutrition might be characterized by the ratio (C-A):org N on a whole-plant basis. This ratio will decrease from about unity to about zero for these plants when going from a situation with nitrate uptake only to a situation with ammonium uptake only. The ratio might increase when nitrogen is depleted, as was shown by Houba et al. [1971] with sugar beet.

4.2 Spinach grown in water-cultures

Results of the water-culture experiment confirmed the expectation that the cation:anion uptake ratio of spinach supplied with NO_3 is close to unity. The calculated slightly higher anion uptake was in accordance with the observation that the solution pH increased. The amount of nitrate assimilated (= organic N) was equal to (C-A) on a whole-plant basis (Figure 4.1). It should be noticed that in the calculations of the present study organic S is neglected as it is only 5% of organic N [Dijkshoorn & van Wijk, 1967]. After NO_3 in the nutrient solution was replaced by SO_4, the NO_3-uptake stopped and the nitrate pool in the plant was assimilated. Sums of cations and anions decreased and cation uptake exceeded anion uptake. This excess cation

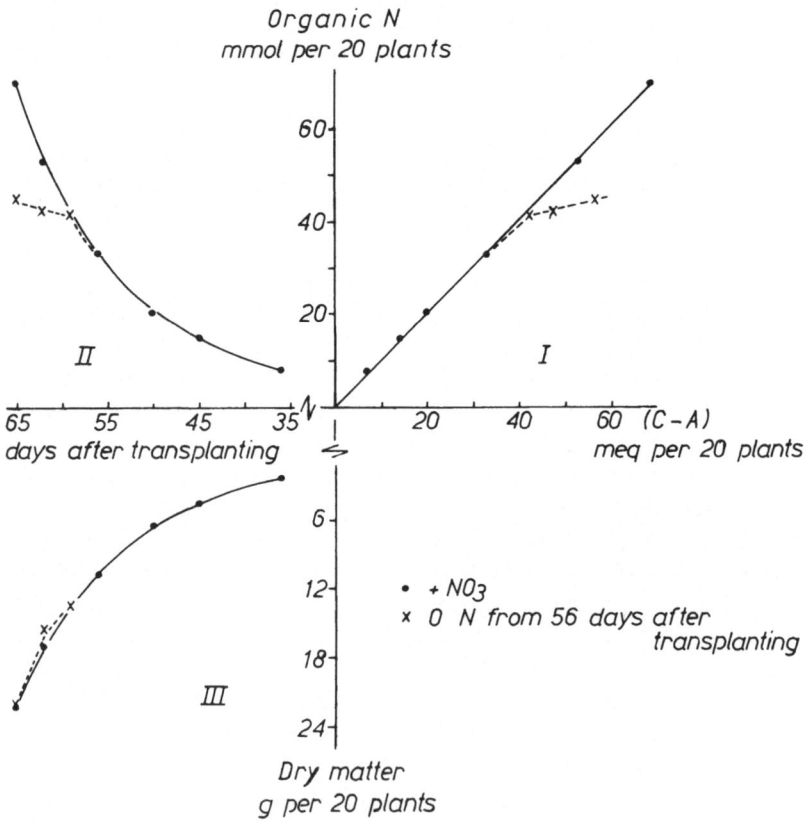

Figure 4.1 Organic-nitrogen production related to (C-A) and dry-matter production of whole spinach plants with continuous and interrupted nitrate supply (Water-culture experiment).

[34]

uptake must have been compensated for by H-ion excretion into the nutrient medium and by the production of organic anions in the root as was suggested by Houba et al. [1971]. As a consequence, carboxylate amounts were no longer equal to the amounts of organic N (Figure 4.1). During nitrogen depletion, (C-A) increased from 32 to 52 meq per 20 plants in shoots and from 1 to 5 meq per 20 plants in roots, while organic nitrogen increased from 30 to 37 mmol per 20 plants in shoots and from 4 to 8 mmol per 20 plants in roots. Obviously the organic anions produced in roots as a result of excess cation uptake were partly transported to the shoots.

The NO_3-contents in aerial parts at final harvest were 497, 224 and 41 mmol per kg dry matter (DM) for spinach grown on NO_3 and without nitrogen for 3 or 9 days, respectively. During the 9-day depletion period, the total dry-matter production of spinach (shoot + root) did not deviate from that of plants continuously supplied with NO_3 (Figure 4.1). However, plants grown without nitrogen for 9 days, at final harvest had root weights amounting to 15% of total dry matter against 11% of total dry matter for plants continuously supplied with NO_3. At final harvest, fresh weights of the aerial parts of plants without nitrogen were about 25% lower than those of the NO_3-plants.

4.3 Spinach grown in sand-cultures with nitrate- and ammonium-nitrogen

In the first experiment, dry-matter production with NH_4-N only was so low (Figure 4.2) that this treatment was deleted in the second experiment. The dry-matter productions per 20 plants per day ranged from 1.3 to 1.7 g in the first (A, spring) and from 0.3 to 0.9 g in the second ($B_1 + B_2$, autumn) experiment, with minor effects of variations in NO_3-N:NH_4-N-ratios in the first and of variations in amounts of N in the second experiment. In the latter, dry matter was significantly lower at the 1:2 ratio. In both experiments shoot:root (DM-basis) and shoot fresh weight were highest for the NO_3-N:NH_4-N = 2:1 ratio.

The NO_3-contents in shoots (Figure 4.3) decreased with increasing age of the plants, the decrease being more pronounced for the lower N-level and higher dry-matter production as pertaining to the first experiment (A). This experiment was carried out in April-May whereas the other was carried out in September-October. In both experiments the NO_3-contents were highest with the NO_3-N:NH_4-N-ratio of 2:1. The NO_3-contents were lowest with the ratios 1:2 (1.5 g N) and 3:0 (2.25 g N). NO_3-contents in roots (not shown) decreased when NO_3-N:NH_4-N decreased for the 1.5 g N treatment; for treatments with 2.25 g N, NO_3-contents were highest for the ratio 2:1. It can be concluded from these results that replacing NO_3-N by NH_4-N is by no means a guarantee for lower NO_3-contents, and that the amount of N, the NO_3-N:NH_4-N-ratio, and other conditions, e.g. light, have to be considered as well.

[35]

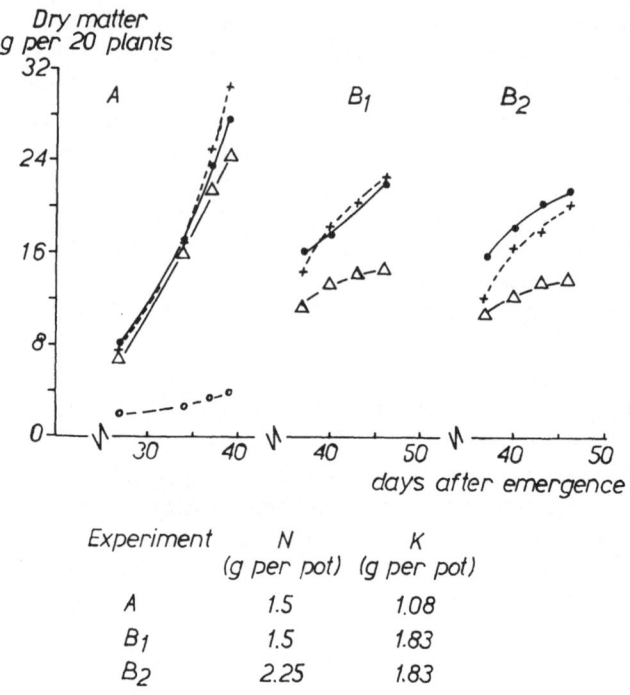

Experiment	N (g per pot)	K (g per pot)
A	1.5	1.08
B₁	1.5	1.83
B₂	2.25	1.83

Figure 4.2 Dry-matter production of whole spinach plants in the sand-culture experiments with varying N- and K-dressings and NO_3-N:NH_4-N-ratios.

If NO_3-N is partly replaced by NH_4-N the chemical composition of spinach is markedly affected as is demonstrated in Table 4 with part of the results of sand-culture B. The total inorganic cation contents (C) in shoots decreased, mainly due to changes in divalent cation contents when the NO_3-N:NH_4-N-ratio decreased. The contents of potassium (the main mono-valent cation) in shoots did not start to decrease until all nitrogen was supplied as NH_4-N (sand-culture A). In roots (C) decreased mainly due to changes in the contents of monovalent cations. The total inorganic anion contents (A) increased in both shoots and roots with changes towards relatively more NH_4-N in the above ratio. These increases were due to increases in SO_4- and H_2PO_4-contents.

The (C-A)-contents in shoots and roots of spinach decreased with lower NO_3-N:NH_4-N-ratios (Free NH_4 was not analysed in the plant material. The contents were estimated to be lower than 50 mmol per kg DM). In Table 4 it is shown that the (C-A)-contents in shoots were balanced for a major part by carboxylates, with oxalate as the one decreasing with higher NH_4-N supply and the other carboxylates remaining more or less constant in the range of NO_3-N:NH_4-N-ratios covered in this table. (Similar results were obtained by Merkel [1975] for spinach on nutrient solutions.) N- and org N-contents in shoots and roots increased in both sand-culture experiments

[36]

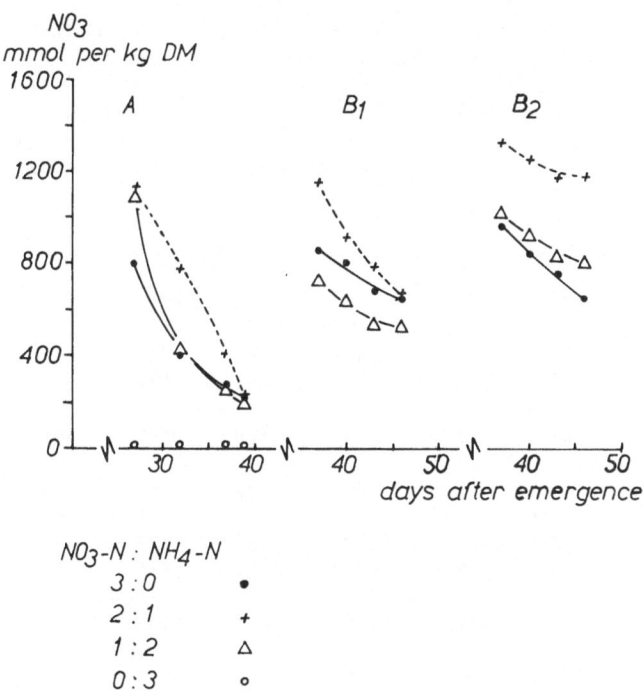

Figure 4.3 Nitrate contents in the aerial parts of spinach in the sand-culture experiments with varying N- and K-dressings and NO_3-N:NH_4-N-ratios.

when the NO_3-N:NH_4-N-ratio decreased (Table 4).

From the relationship between organic nitrogen and (C-A) (Figure 4.4), it can be observed that the ratio (C-A):org N (on a whole-plant basis) was about one with NO_3-N only and that it decreased to about zero when only NH_4-N was applied, as suggested by van Egmond [1978]. For spinach, the (C-A):org N-ratio indicates fairly well the extent of NO_3- or NH_4-nutrition. As the contribution of the roots to (C-A) and org N is relatively small, these indications could also be found with the use of aerial parts only.

As the carboxylate pool can also be affected by the supply of plant nutrients other than nitrogen, the effects of K-supply were studied in more detail. Comparable (C-A):org N-ratios were found in the first and second experiment (Figure 4.4), although K-supply per pot differed by 0.75 g K. If 1.83 g instead of 1.08 g K were given, the K-contents in shoots of spinach increased by about 800 meq per kg DM in sand-culture B as is shown in Table 5 with a part of the results of this experiment. The lower K-contents were partly compensated for by higher Na-, Mg- and Ca-contents, resulting in (C)-contents which were only 200-300 meq per kg DM lower. However, with lower K-supply the (A)-contents in the shoots were also about 200 meq per kg DM lower, mainly because of the lower NO_3-contents (Table 5). As a result of these changes, the (C-A)-contents in shoots were hardly affected by K-sup-

Table 4. *Yield and chemical composition of spinach grown in sand with per pot added 1,5 g N, as NO_3- and/or NH_4-N, and 183 gK (Sand-culture experiment B, third harvest).*

NO_3-N: NH_4-N	3 : 0		2 : 1		1 : 2	
	Shoot	*Root*	*Shoot*	*Root*	*Shoot*	*Root*
Fresh weight[a])	231		261		148	
Dry weight	17.7	2.7	18.2	2.4	12.3	1.9
K[b])	2124	679	2175	875	1940	1188
Na	234	353	157	587	68	143
Mg	597	844	414	588	222	494
Ca	970	576	386	461	184	476
Σ C	3925	2452	3132	2508	2414	2301
CI	38	36	56	30	65	27
SO_4	56	142	150	356	333	564
H_2PO_4	194	297	289	358	375	441
NO_3	675	450	799	287	534	85
Σ A	963	925	1294	1031	1307	1117
N	3787	2740	4593	2873	5255	3443
org N	3112	2290	3793	2586	4721	3358
(C - A)	2963	1527	1837	1477	1107	1184
Organic anions						
Succinate	31		31		28	
Fumarate	22		16		14	
Malate	230		170		213	
Oxalate	2354		1324		537	
Σ	2637		1541		792	
(C - A) - Σ Organic anions	326		296		315	

a) g per 20 plants.
b) meq per kg DM.

ply. In roots (C)-, (A)- and (C-A)-contents were hardly affected by a reduced K-supply; with high N-supply, (C) and (C-A) were somewhat lower.
The N- and org N-contents in shoots and roots tended to be lower values
with low K-supply, except when N-supply was high. Dry matter and fresh
weights of shoots were about 10 and 25% lower for plants with 1.08 g K than
for the ones with 1.83 g K (Table 5).

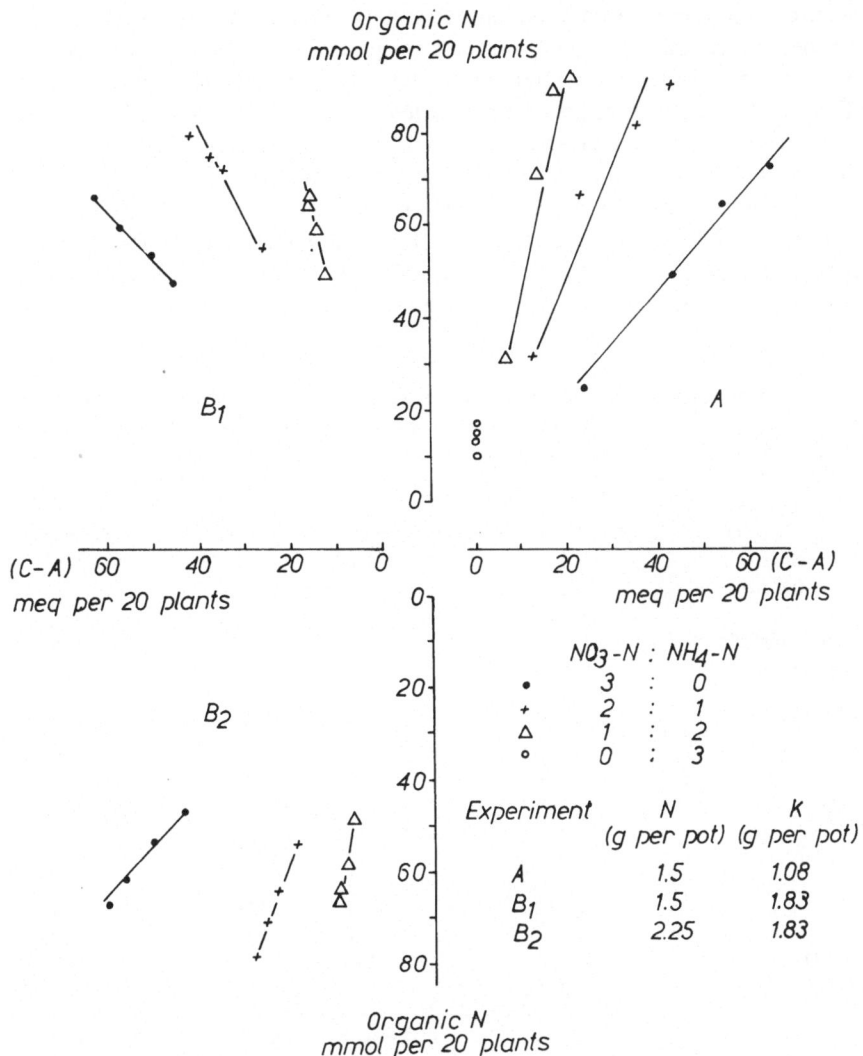

Figure 4.4 Relationship between organic nitrogen and (C-A) of whole spinach plants in the sand-culture experiments with varying N- and K-dressings and NO3-N:NH4-N-ratios.

4.4 The effects of nitrate-nitrogen on yield and chemical characteristics of spinach grown on various soils

4.4.1 The amount of nitrate-nitrogen applied

The dry-matter yields of spinach in the experiments covered in Figure 4.5 increased up to $1.5\,g\ NO_3\text{-}N$ per pot, which level frequently brings about maximum spinach yields in pot experiments under the conditions described in section 3.1.3. NO_3-contents in spinach were found to be strongly affected by nitrate concentrations in the soil solutions. The differences in Figure 4.5 must furthermore be attributed to plant age, light conditions, soil type, etc. It can be seen that in general beyond 1.0 g NO_3-N per pot sharp increases in NO_3-contents are to be expected.

A representative example of the relationship between NO_3-N concentration in soil after harvest and NO_3-content in the plant is shown in Figure 4.6 where it can be seen that relatively high NO_3-contents in spinach occur above a concentration of about 5 mg NO_3-N per kg soil. Below this level in

Table 5. *Yield and chemical composition of the aerial parts of spinach grown in sand, as influenced by variations in quantities of NO_3 and K applied (Sand-culture experiment B, third harvest).*

NO_3-N - dressing (g per pot)	1.5		2.25	
K dressing (g per pot)	1.08	1.83	1.08	1.83
Fresh weight[a]	176	231	159	220
Dry weight	16.0	17.7	14.6	17.7
K[b]	1321	2124	1456	2153
Na	322	234	286	210
Mg	618	597	569	504
Ca	1354	970	1527	1161
Σ C	3615	3925	3838	4028
Cl	42	38	41	44
SO_4	48	56	61	110
H_2PO_4	195	194	224	199
NO_3	498	675	626	764
Σ A	783	963	952	1117
N	3654	3787	3909	3846
org N	3156	3112	3282	3082
(C - A)	2832	2963	2886	2911
Organic anions				
Succinate	27	31	42	48
Fumarate	10	22	29	31
Malate	172	230	128	154
Oxalate	2391	2354	2477	2351
Σ	2600	2637	2676	2584
(C - A) - Σ Organic anions	232	326	210	327

[a] g per 20 plants.
[b] meq per kg DM.

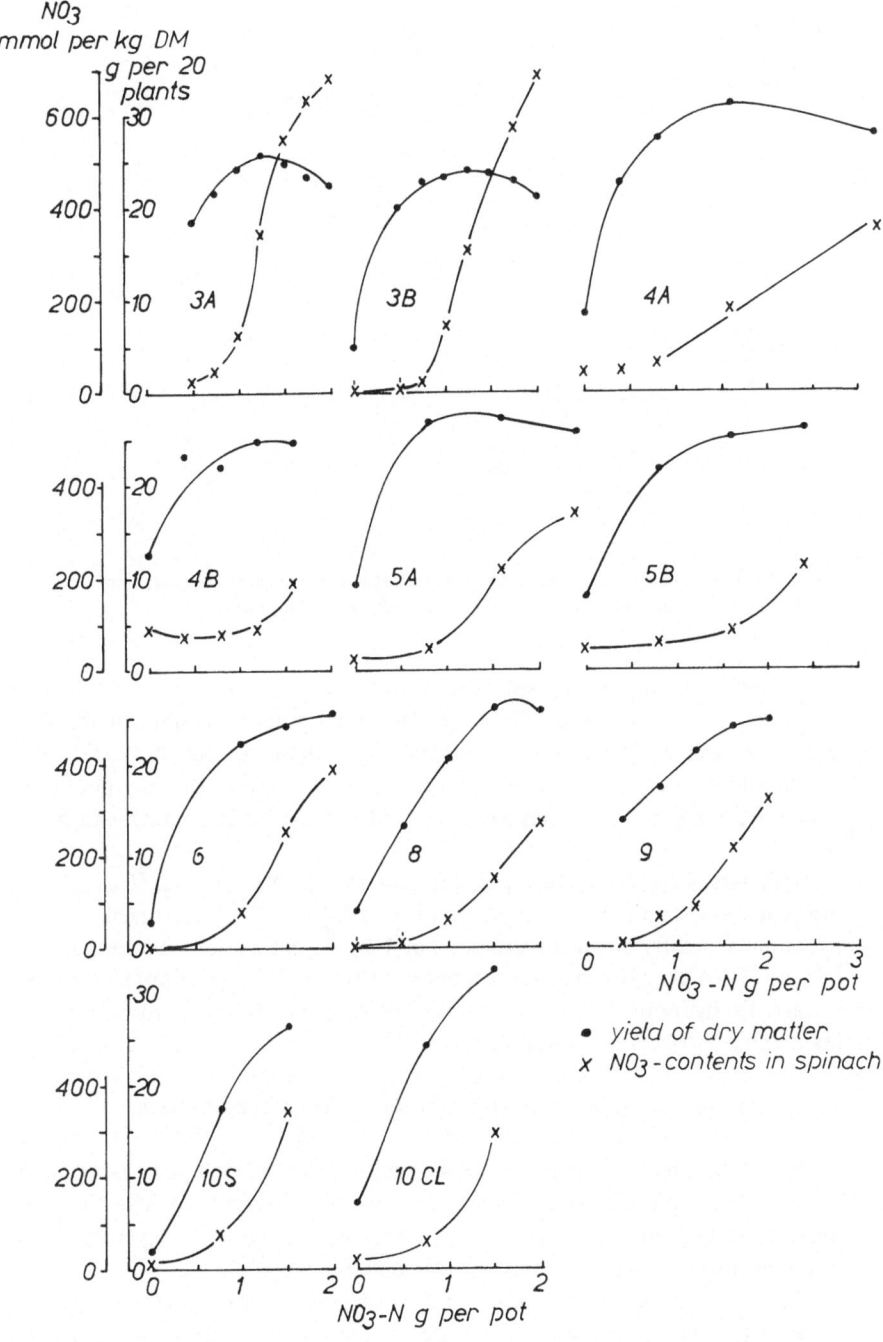

Figure 4.5 Effect of variations in nitrate dressing on NO₃-contents and yields (DM) of spinach in experiments 3, 4 (moisture at 60% WHC), 5, 6, 8, 9 (pH-KCl=5.5) and 10 (without Mo-spraying, S=sand, CL=clay loam).

[41]

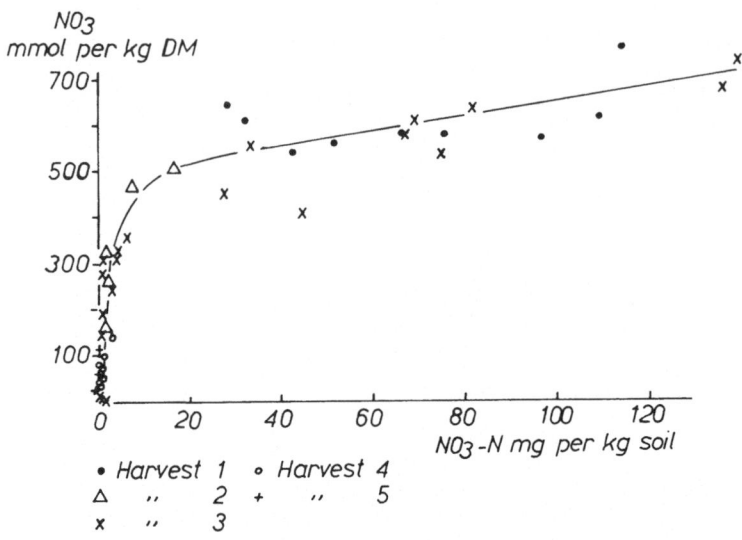

Figure 4.6 Relationship between NO3-concentration in soil after harvest and NO3-content in the aerial parts of spinach (Experiment 3B, individual data).

the range of 0-5 mg per kg soil, the NO_3-contents in plant material decreased rapidly to almost zero. These results were observed in spinach grown in spring as well as in autumn. It is clear from these results that only when the supply of NO_3-N in the pots is practically exhausted (0-5 mg per kg soil equals 0-35 mg per pot) relatively low NO_3-contents in spinach can be expected.

With increasing NO_3-N-supply, the contents of Na, Ca, Mg, (C-A), N and org N increased and the contents of Cl, H_2PO_4 and SO_4 decreased with the K-contents mostly being unaltered. For NO_3-N-applications exceeding 0.8 g per pot, (C-A):org N-ratios ranged between 0.9 and 1.1. For pots with lower nitrate applications these ratios were between 1.1 and 1.9, thus reflecting instances of progressive N-deficiency.

4.4.2 Distribution of plant nutrients in the aerial parts of spinach

As can be observed from Figure 4.7, within one spinach plant large differences in NO_3-contents existed between leaves of different age. The NO_3-contents increased with leaf age, the contents in old leaves being higher than in younger ones regardless of the nitrate dressing (Figure 4.7).

The dry weights and the weight ratios of laminae and petioles for pairs of consecutive leaves increased with age of the leaves. Within each pair, this ratio remained more or less constant during the growth period, with a value of about 5 for the oldest and of about 2 for the youngest leaves.

The NO_3-contents in the petioles of the high-NO_3 plants were 3 to 4 times higher than in the laminae (Figure 4.7). At the low NO_3-level, the nitrogen

[42]

was depleted about 32 days after emergence and from then on the NO_3-contents decreased to very low values in both petioles and laminae.

For the high-NO_3 treatment, the total amount of NO_3 calculated per leaf-pair increased during the growth period. For the low-NO_3 treatment this amount decreased in all leaf pairs from the 32nd day after emergence onwards. For the high and the low NO_3-supply, 50 and 75% of the total amounts of NO_3 per plant, respectively, were found in the oldest four leaves. In this experiment no evidence was found of redistribution of nitrate in spinach.

In contrast to NO_3-contents, the N-, org N- and H_2PO_4-contents decreased with increasing leaf age. In the laminae, the org N-contents were 2 to 3 times higher than in the petioles. The other plant nutrients will not receive detailed attention here. Variations in org N- and (C-A)-contents in laminae and petioles of different age in spinach, as shown in Figure 4.8 with the results obtained at one of the harvests, were found to be very similar to those observed by van Egmond [1971] for sugar beet. In 1975, this author deduced that carboxylates were not redistributed in sugar beet. If it is assumed that the same holds for spinach, it is obvious that, as carboxylates in spinach are mainly formed during the process of NO_3-reduction and the distribution of carboxylates (C-A) more or less runs counter to that of org N (Figure 4.8), org N must have been redistributed. Although the (C-A)- and org N-contents varied for leaves of different ages, the (C-A):org N-ratio was about one for the aerial parts of plants with a high nitrate supply. With a low nitrate supply, the same was true until NO_3 became depleted.

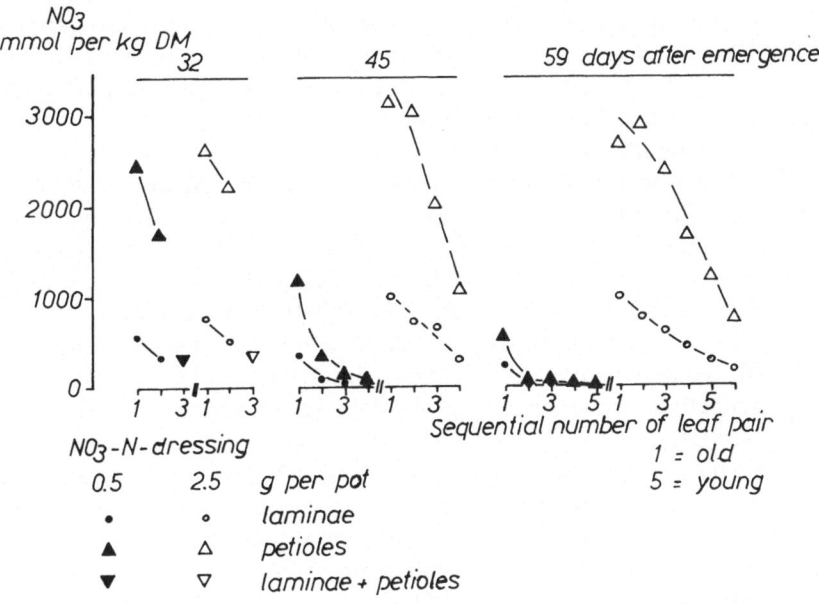

Figure 4.7 NO_3-contents in laminae and petioles of spinach leaves of different age for two levels of NO_3-supply (Experiment 1).

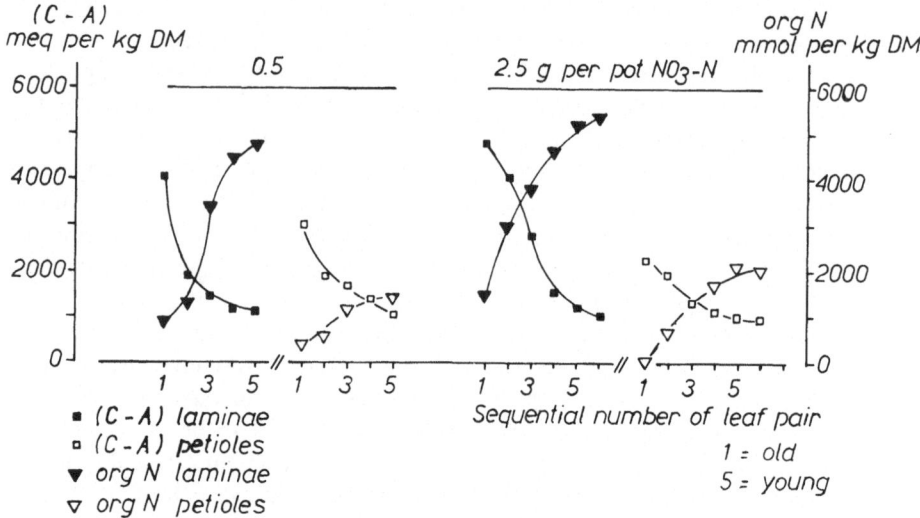

Figure 4.8 (C-A) and org N-contents in laminae and petioles of spinach leaves of different age at two levels of NO3-supply, 59 days after emergence (Experiment 1).

4.4.3 Age of plants and season of growth

Plant age and nitrate supply (section 4.4.1) had a marked effect on the NO_3-contents in spinach (Figure 4.9). The NO_3-contents decreased after a certain age; this age is related to the amount of NO_3-N applied and to the stage of dry-matter production. It is evident that with a high NO_3-N supply a high NO_3-content will be maintained for a longer period.

From the results of experiments 3A and B in Figure 4.9 it can be seen that the NO_3-contents in the spinach grown in spring (B) were lower than those found in autumn (A), and that the reverse was true for dry-matter production. The differences were mainly due to the different light conditions. The average daily radiations (see also Figure 3.2) during the growth periods in 1979 (September-November) and in 1980 (March-May) were 7.5 and 13.5 MJ per m^2, respectively. In autumn radiation decreased and in spring it increased during the growth period. These differences will have affected both NO_3-assimilation and dry-matter production. In the period from 29 to 56 days after emergence in 1979 (A), the mean dry-matter productions were 0.5, 0.6 and 0.6 g per day with 1.0, 1.25 and 1.50 g NO_3-N, respectively, whereas in 1980 (B) from 37 to 53 days after emergence 1.4, 1.5 and 1.5 g per day were produced. As can be expected from the dry-matter productions, the N- and org N-contents in the spring experiment (B) decreased to lower levels than in the autumn experiment (A).

[44]

4.4.4 Time of application of nitrate-nitrogen

A dressing of 1.0 g per pot N, applied before sowing, resulted in fresh weights which were about 10% lower than those obtained with higher N-dressings 49 and 56 days after emergence in experiment 3A. The dry weights, however, did not show significant differences (Table 6).

The NO_3-contents were affected by variations in the quantity of NO_3-N applied, but not by variations in the timing of application (Table 6). The results obtained in experiment 3B were similar to those of experiment 3A presented in Table 6.

4.4.5 Soil-moisture content

The effects of variations in moisture content of the soil on the dry-matter yields and NO_3-contents are shown in Table 7 with data from experiment 4B. The higher NO_3-contents and lower dry-matter yields with low soil-moisture content (40%) are in accordance with earlier data presented in the literature (section 2.6.3). The levels of 60 and 80% of maximum water-holding capacity (60% is about pF 2 for Dutch soils) did not differ in their effects on dry-matter yields and NO_3-contents of spinach (Table 7). Similar conclusions can be drawn from the results of experiment 4A.

4.4.6 Soil pH

The fresh- and dry weights of spinach were not affected by variations in pH-levels as used in experiment 9. The NO_3-contents and the N-amounts in the aerial parts of spinach, however, differed considerably (Figure 4.10), especially for the intermediate NO_3-N-levels. It should be noted that with pH-KCl-levels of 5.5 and 6.0 the relationship between NO_3-dressings and NO_3-contents was the same as shown in section 4.4.1 (Figure 4.5). The NO_3-contents at these pH-levels were lower than those at pH 4.5 and 5.0

As higher pH-levels are more favourable for the growth of micro-organisms, NO_3-immobilization might have been somewhat higher in soil with these higher pH-levels, resulting in lower NO_3-contents in spinach. Moreover, at these higher pH-levels NO_3-uptake might have been suppressed, as was also observed by Rao & Rains [1976]. A pH-KCl-value of 5.5 for the soil used in this experiment corresponds with a pH-H_2O-value of about 6.0.

The acidic-uptake pattern resulting from NO_3-nutrition gave rise in the soils to increases in pH-KCl of 0.2-0.3 units for all NO_3-dressings, except 0.4 g N per pot. The pH-range for most N-dressings must therefore be considered to have been wider than suggested in the experimental design.

4.4.7 Mo-spraying on two soil types

Mo-spraying did not have any effect on the NO_3-contents, the dry- and fresh weights and on other characteristics of spinach (experiment 10).

[45]

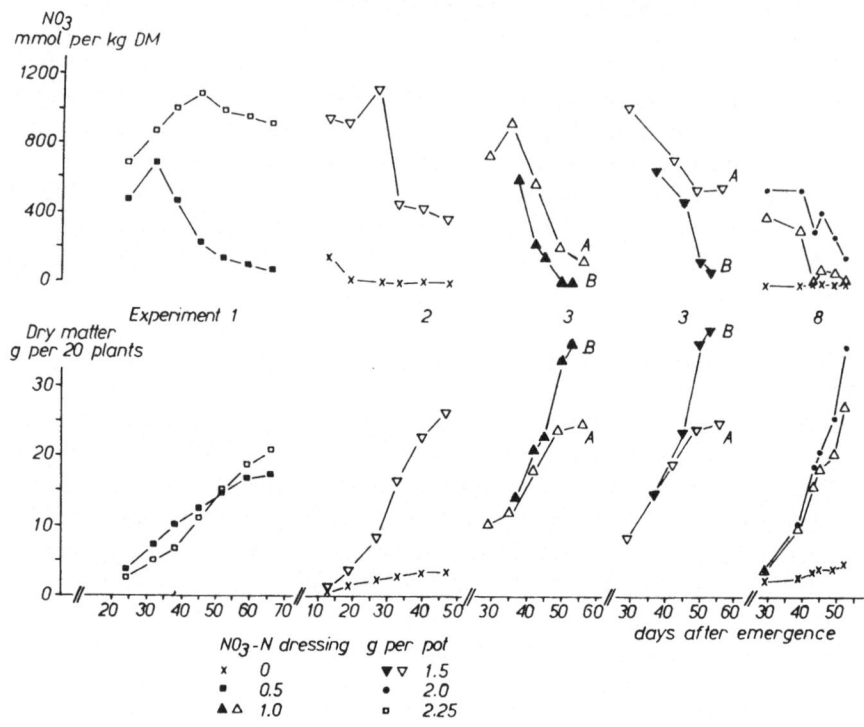

Figure 4.9 Dry-matter production and NO3-contents in the aerial parts of spinach plants of different age and with varying NO3-N dressings.

The NO_3-contents in plants grown on the sandy soil were higher than of those grown on the clay loam, with yields being higher for the latter soil (Figure 4.5). With equal N-dressings per pot, the NO_3-concentration in the soil solution will have been higher in the sandy soil, thus enhancing NO_3-uptake by young plants and creating higher NO_3-contents in the older leaves (section 4.4.2). As upon exhaustion of available soil N, in these older leaves NO_3 is less easily reduced, somewhat higher NO_3-contents might be expected in leaves of plants grown on sandy soils.

Table 6. *Effects of variation in the quantity and in the timing of NO_3 application on the NO_3-contents and dry-matter yields of spinach (Experiment 3A).*

NO3-N-dressing (g per pot)	1.0			1.25		1.5
		+ 0.25[a])	+ 0.25[a]) + 0.25[b])		+ 0.25[b])	
Days after emergence	NO3	(m mol per kg DM)				
42	566	849	–	897	–	717
49	205	510	533	485	767	535
56	127	333	588	346	460	552
	Dry matter (g per 20 plants)					
42	18.3	18.3	–	18.1	–	20.1
49	24.1	23.8	23.8	24.7	25.0	24.6
56	25.3	25.6	25.9	26.7	27.1	25.4

[a]) Applied 29 days after emergence.
[b]) Applied 42 days after emergence.

Table 7. *Effect of variations in soil-moisture contents and NO_3-N dressings on the NO_3-contents and dry-matter yields of spinach (Experiment 4B).*

NO3-N-dressing (g per pot)	0.8	1.2	1.6
Soil-moisture content (% of maximum waterholding capacity)	NO3 (m mol per kg DM)		
40	158	227	283
60	82	96	188
80	43	47	156
	Dry matter (g per 20 plants)		
40	22.2	21.1	19.2
60	22.9	25.2	25.3
80	21.6	23.8	28.8

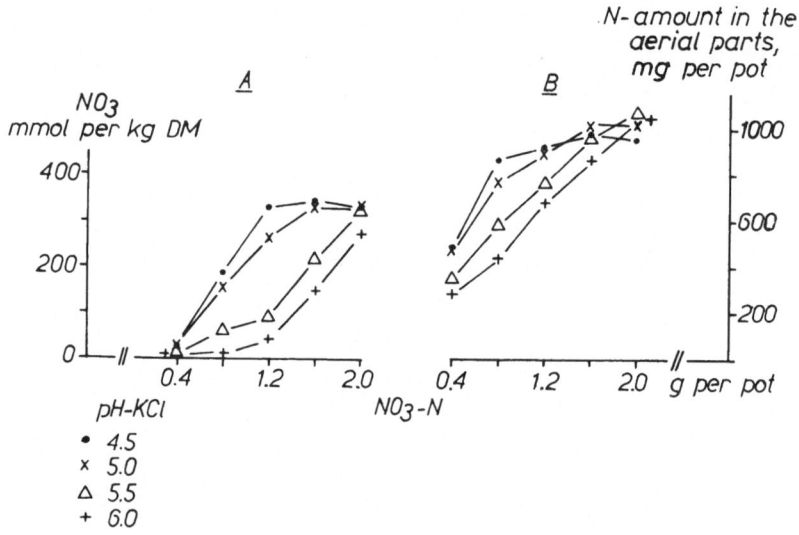

Figure 4.10 Effects of variations in soil pH on (A) NO₃-content and (B) N-amount in the aerial parts of spinach supplied with different quantities of NO₃-N (Experiment 9).

4.5 The effects of variations in N-form in soil cultures on yield and chemical characteristics of spinach

4.5.1 Ammonium- and nitrate-nitrogen dressings

Since normally in soil NH_4 is rapidly transformed to NO_3, the effects of ammonium- or nitrate-nitrogen in soil-culture can only be tested if a so-called nitrification inhibitor is used.

Without a nitrification inhibitor, as in experiment 6, NH_4-dressings resulted in soil pH decreases of 1.0 to 1.5 units because of initial NH_4-uptake and subsequent nitrification of non-absorbed NH_4 (Table 8). The low pH-values caused yields to decline and, as nitrification was not suppressed, the NO_3-contents in spinach sometimes were even higher than with NO_3-dressings. Addition of a nitrification inhibitor, in the form of DCD, to pots with nitrate-nitrogen did not have any effect on yield or NO_3-contents of spinach. With ammonium dressings, however, DCD suppressed the NO_3-contents considerably (Table 8). The unfavourable effect on soil pH again manifested itself in lower yields.

The effects of NH_4-N on the ratio (C-A):org N as discussed in section 4.3 were confirmed in this experiment. With ratios of 0.5 or less, lower NO_3-contents were found, but in that situation yield was also substantially suppressed.

[48]

Table 8. *Nitrate contents and dry-matter yields of spinach, and soil pH values after harvest as affected by ammonium- and nitrate-nitrogen applications. (Experiment 6).*

	Without DCD				With DCD[a]) (30 mg per pot)		
N-dressing (g per pot)	0	1.0	1.5	2.0	1.0	1.5	2.0
N-source							
NO_3 (m mol per kg DM)	3						
$(NH_4)_2SO_4$		44	362	405	19	67	116
$Ca(NO_3)_2$		81	254	387	76	242	395
Dry matter (g per 20 plants)	3.1						
$(NH_4)_2SO_4$		21.1	19.8	19.7	20.1	22.5	22.4
$Ca(NO_3)_2$		22.6	23.8	26.0	22.4	24.7	25.1
Soil pH-KCl (after the harvest)	5.5						
$(NH_4)_2SO_4$		4.5	4.1	4.0	4.6	4.3	4.2
$Ca(NO_3)_2$		5.4	5.6	5.6	5.5	5.6	5.6

a) DCD = dicyandiamide, a nitrification inhibitor.

Under the conditions prevailing in pot cultures (restricted volume, no leaching, etc.), the effects of nitrification inhibitors on the NO_3-concentration in incubated soil samples were very convincing. Quantities amounting to 93, 50, 27, 34 and 5% of the total amount of inorganic nitrogen (dressing 1.5 g N per pot) were available as NO_3 after a 65-day incubation for NH_4-N without inhibitor, NH_4-N with 30 and 90 mg DCD per pot and NH_4-N with 30 and 90 mg nitrapyrin per pot, respectively (Experiment 6). In the same order, spinach yields were reduced by 16, 10, 26, 33 and 74% compared with the yield of a NO_3-dressed crop. These reductions can be partly ascribed to the effects of nitrification on soil pH, as was mentioned above, and partly to an enhanced NH_4-uptake (section 4.3).

4.5.2 Light intensity- and temperature effects with different amounts of nitrogen and varying NO_3:NH_4-ratios

Variations in quantity of N-dressing and in NO_3-N:NH_4-N-ratio did not affect dry-matter yields in experiment 7 (Table 9). The NO_3-contents, however, were significantly affected by variations in these factors and in light intensity and temperature in the growth chambers which contained the spinach plants (Table 9, Figure 4.11).

[49]

Table 9. *Nitrate contents, (C - A): org N values and dry-matter yields of spinach as affected by variations in N-dressings, NO_3: NH_4-ratio and plant age (Experiment 7)[a]).*

N-dressing (g per pot)	1.5			2.25		
NO_3-N	10	7	4	10	7	4
NH_4-N	0	3	6	0	3	6
Days after emergence						
NO_3 (m mol per kg DM)						
27	668	607	425	648	611	549
37	510	491	243	590	639	410
(C - A): org N						
27	0.9	0.7	0.7	0.9	0.7	0.7
37	0.8	0.6	0.5	0.8	0.6	0.5
Dry matter (g per 20 plants)						
27	10.8	11.0	10.2	10.7	10.5	10.5
37	21.0	21.4	18.8	20.7	20.6	20.2

[a]) Temperature $17°C$; light intensity (400-700 nm): 62 W per m^2 for first harvest and 70 W per m^2 for second harvest.

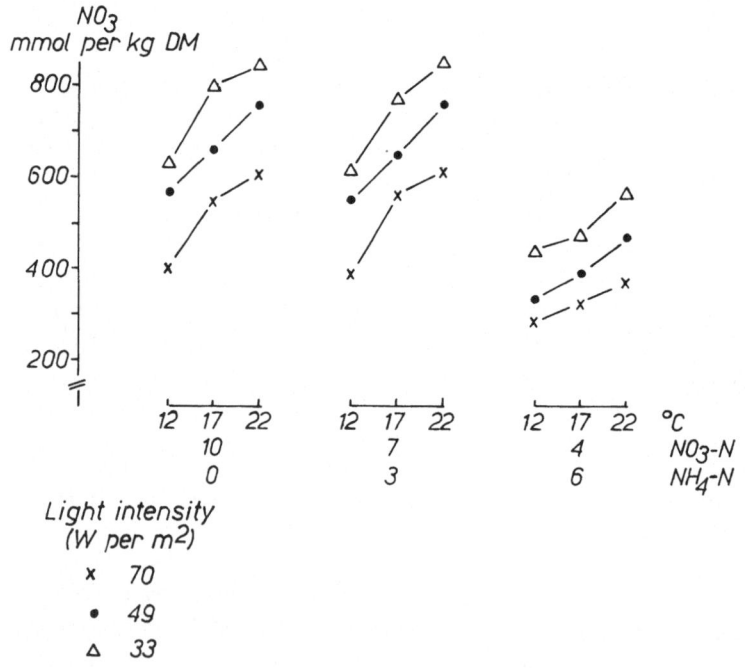

Figure 4.11 Effect of variations in NO_3:NH_4-ratio, temperature and light intensity on NO_3-contents in spinach (means of two N-dressings). (Experiment 7, final harvest).

[50]

Higher temperature and lower light intensity (characteristics of a greenhouse in the winter season) gave rise to higher NO_3-contents, with the lowest values for the NO_3-N:NH_4-N-ratio considered most favourable in this respect. It should be noticed that in this experiment the spinach plants were harvested at the beginning of the photoperiod. However, at these relatively high NO_3-contents the effects of the diurnal variation in light on the NO_3-contents, as observed by e.g. Cantliffe [1972-2], could be expected to be of minor importance.

The suppressive effects of increases in N-dressing and of decreases in the NO_3-N:NH_4-N-ratio on dry-matter yield and NO_3-content of leaves, respectively, were smaller than expected on the basis of the results obtained in experiments with media containing more sand.

There is evidence that the clay loam soil used in experiment 7, with a CEC (= cation exchange capacity) of 20 meq per 100 g soil, has a capacity to fix NH_4-ions (and K-ions)(J.H.G. Slangen, private communication). This implies that at least a portion of the NH_4-N will have been removed from the soil solution, thus decreasing the NH_4-concentration in the rhizosphere and the uptake of NH_4. This reduction in available N, to be expected for the lower NO_3-N:NH_4-N-ratios, will have manifested itself mainly in a reduction in quantity of available NO_3. The data of the NO_3-contents and (C-A):org N-values, shown in Table 9, were in accordance with these assumptions.

For the somewhat lower (C-A):org N-values in the second harvest no explanation can be given.

4.6 Comparison of the performances of different N-carriers as reflected in yields and NO_3-contents of spinach

The results of experiment 5A without inhibition of nitrification (Figure 4.12) showed that spinach yields were somewhat higher with Peraform and sulphur-coated urea (SCU) than with $Ca(NO_3)_2$ (the yields and NO_3-contents obtained with the use of sewage sludge were low and will not be discussed). Damage from high salt concentrations with $Ca(NO_3)_2$, mainly at the highest N-dressing, account for the relatively low yield obtained with this N-carrier. Irrespective of the quantity of N applied the NO_3-contents were always about 100 mmol per kg DM lower with SCU than with $Ca(NO_3)_2$ and Peraform.

Inhibition of nitrification (experiment 5B) resulted in substantial yield reduction for those carriers which provide N as NH_4. Judging from the yields obtained with $Ca(NO_3)_2$ in both experiments, no effect of inhibition other than that on nitrification was found in experiment 5B. The low NO_3-contents obtained with the high rate of N supplied as NH_4 (Figure 4.12) also testify to the effectiveness of the inhibition in this experiment.

[51]

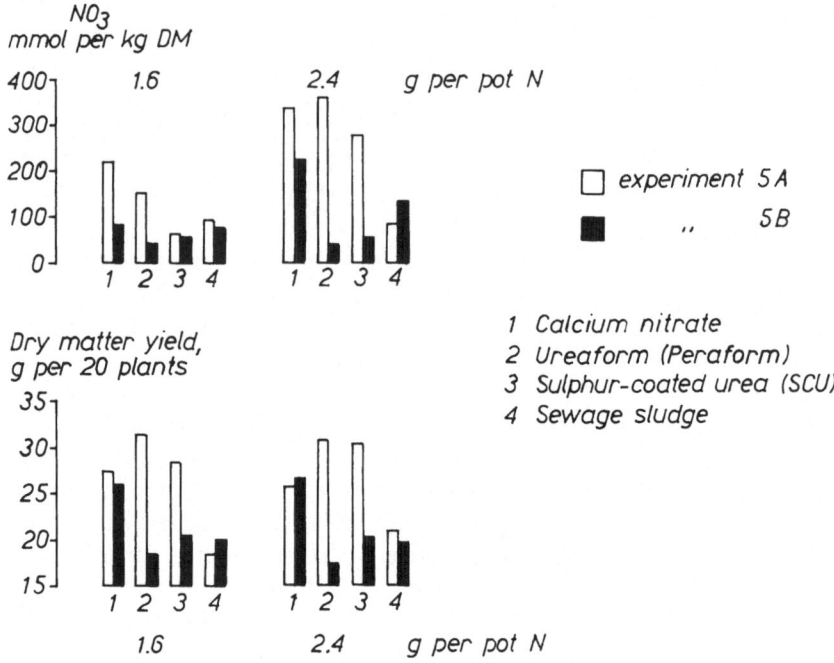

Figure 4.12 Effects of two dressings of N-carriers on the NO3-contents and dry-matter yields of spinach (For SCU and sewage sludge the total amounts of N added per pot were three times higher than the quantities of N indicated in the figure)(Experiment 5).

With $CaCN_2$ as N-carrier in experiment 6, dry-matter yields were less than 20 g per 20 plants and NO_3-contents less than 120 mmol per kg DM for N-dressings up to 2.0 g per pot. (C-A):org N-values being lower than 0.4 manifested the extent of NH_4-uptake and -assimilation. With pH-KCl-values as high as 6.5, resulting from the decomposition of cyandiamide into urea, DCD and NH_4, ammonia-toxicity [Schenk & Wehrmann, 1979] is highly probable and accounts for the low yields.

4.7 The effects of N in manures on yield and chemical composition of spinach

The nitrogen from the manures used in experiment 8 was only partly available for uptake by plants. Available N, as calculated from the results of an incubation study, was therefore used to construct Figure 4.13 B and C. In these figures, one graph could be drawn for both the high and low amounts of pig-manure slurry (PMS) and one graph for the high and low amounts of farmyard manure (FYM).

With both FYM and PMS, NO_3-contents were lower and dry-matter yields were higher than obtained with $Ca(NO_3)_2$. From this result, it might be de-

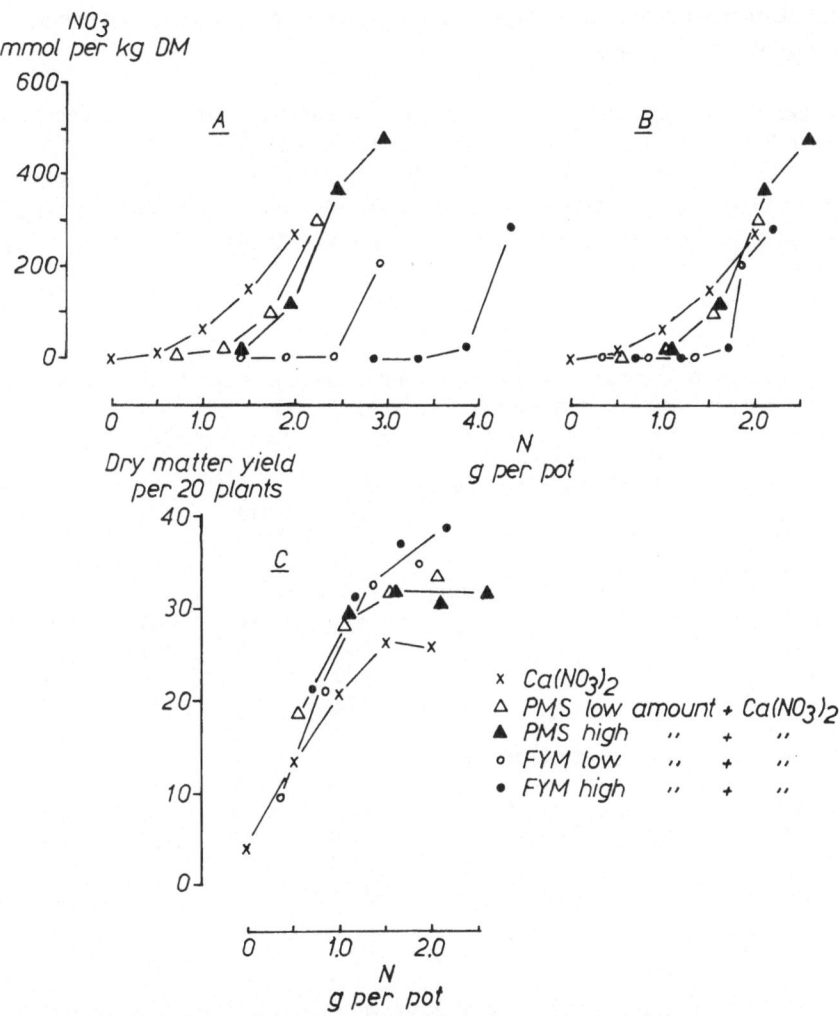

Figure 4.13 Effects of N applied as calcium nitrate, as pig-manure slurry (PMS) + calcium nitrate or as farmyard manure (FYM) + calcium nitrate on the NO_3-contents and dry-matter yields of spinach
(Experiment 8).
A. On horizontal axes, the total amounts of N applied are shown.
B+C. As in A but with 75 and 25% of the total N applied as pig manure slurry and farmyard manure, respectively.

duced that these fertilizers are promising for use in spinach growing. Nevertheless, it is questionable whether under field conditions the mineralization of organic N from these materials coincides with the growth period of spinach, and whether the quantities mineralized are sufficiently independent of weather conditions to always guarantee an adequate supply of available N.

4.8 Effects of variations in P-dressing and in soil P-status on yield and chemical composition of spinach

On the average there was a tendency of the NO_3-contents to decrease when P-dressings and values of available soil P increased (Table 10).

In this experiment with relatively high available soil P-levels, dry-matter yields were found to be not affected by variations in size of the P-dressings, but at the highest soil P-level yields were about 10% higher than at all other levels.

Table 10. *Effects of variations in P-dressings and in values of available soil P on NO_3-contents and dry-matter yields of spinach (Experiment 11).*

P-dressing (g P per pot)	Available soil P (P-AL values, mg P per kg soil)	157	214	284	380
	NO_3 (m mol per kg DM)				
0		474	454	409	450
0.22		536	380	475	4i1
0.44		429	390	403	398
0.66		538	397	352	373
	Dry matter (g per 20 plants)				
0		24.9	24.9	26.3	27.9
0.22		24.5	24.9	24.2	27.5
0.44		26.0	25.7	24.6	28.1
0.66		25.0	24.9	24.7	28.6

4.9 Effects of variations in size of K-dressings and in type of K-carrier on yield and chemical composition of spinach

From Figure 4.14 it can be seen that spinach responded in yield to K-dressings up to 30% yield increase obtained with the highest dressing. The differences in response to KCl and K_2SO_4 were small and statistically not significant. With the use of K_2SO_4, NO_3-contents increased by about 250 mmol per kg DM (between 0 and 1.66 g K per pot) in autumn and by about 100 mmol per kg DM in the spring experiments. The differences in NO_3-contents between KCl-treated and K_2SO_4-treated plants were about 100 mmol per kg DM (Figure 4.14).

The uptake of NO_3 was likely to be enhanced by high K-uptake but due to competition between NO_3 and Cl for attachment to absorption sites in plant roots, lower NO_3-contents were found in KCl-treated than in K_2SO_4-treated plants.

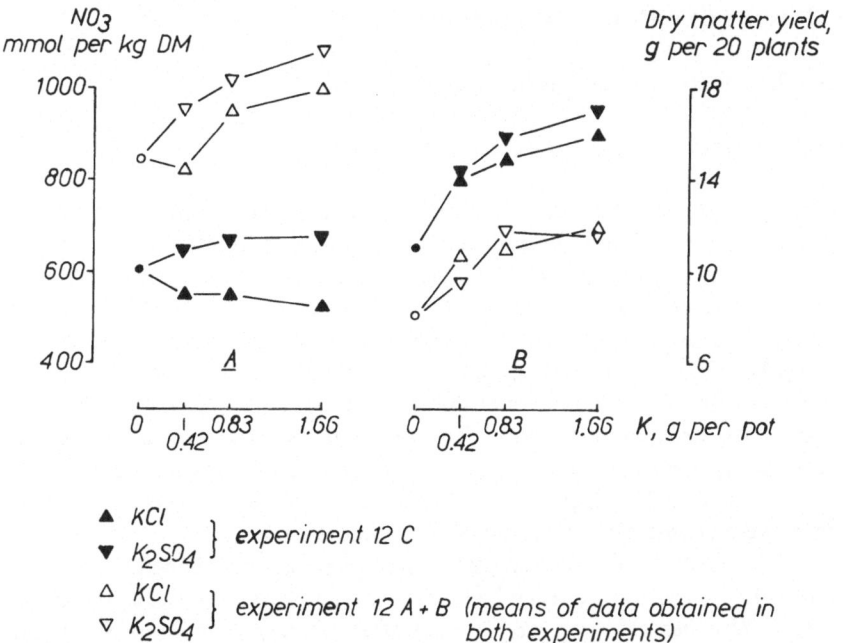

▲ KCl
▼ K₂SO₄ } experiment 12 C

△ KCl
▽ K₂SO₄ } experiment 12 A + B (means of data obtained in both experiments)

Figure 4.14 Effects of variations in the amount of K applied and the type of K-carriers on (A) NO$_3$-contents and (B) dry-matter yields of spinach. (Data are means of values obtained on three soils) (Experiment 12).

4.10 Effect of varietal differences on yields and NO$_3$-contents in spinach

At both harvests, NO$_3$-contents of the variety Medania were significantly higher than those of the other varieties, while dry-matter yields of Medania were significantly lower (Table 11). The NO$_3$-uptake- and -reduction patterns for Nobel, Virkade and Califlay must have been similar, whereas these patterns for Medania differed from those of the other varieties.

It should be remarked that the plant- to -plant variation in NO$_3$-content was larger than normally found for other nutrients. Within one variety and one harvest, NO$_3$-contents may differ among plants by a factor 10.

Table 11. *Effect of plant age and varietal differences in NO$_3$-contents and dry-matter yield of spinach (Experiment 13).*

Variety	Nobel	Virkade	Medania	Califlay
Days after planting				
	NO$_3$ (m mol per kg DM)			
52	336	253	608	255
59	202	106	501	130
	Dry matter (g per plant)			
52	5.1	5.9	4.0	5.4
59	7.2	9.2	6.0	7.8

[55]

5 Results of outdoor experiments

5.1 The effects of variations in nitrate dressings

5.1.1 Yields and dry-matter contents of spinach

In all but one (i.e. exp. 7) experiments, compared with no nitrogen, fresh yields were more than doubled with applications of 50 or 75 kg NO_3-N per ha (Table 12, part a). Yield increases diminished with higher dressings; but dressings of 150 and 200 kg NO_3-N per ha gave invariably the highest fresh yields. Because of rainfall during or just before harvesting, in experiments 1B (final harvest), 4, 6 and 7, fresh yields were somewhat overestimated and dry-matter contents (Table 12, part c) somewhat underestimated. In all trials, dry-matter contents decreased with increasing nitrate dressings; those in experiment 3 were relatively low because of the young age of this crop at harvest. The high dry-matter contents in 1B (31 days after emergence) were caused by dry weather shortly before harvest.

With respect to dry-matter yield, the trials can be divided into three groups. In one group of trials highest yields per harvest were in the range of 150-175 g per m^2 (exp. 1B, 31 days after emergence and exp. 3, 24 and 26 days after emergence), in a second group the highest yields ranged from 200 to 250 g per m^2 (exp. 1B, 37 days after emergence, exps. 2B and 4 and exp. 3, 30 days after emergence) and in a third group the highest yields ranged from 290 to 350 g per m^2 (exp. 1B, 44 days after emergence and exps. 5, 6 and 7) (Table 12, part d). Dry-matter yields were closely correlated with fresh yields. The decreases in yield obtained with the highest nitrate dressings were caused by high salt concentrations in the soil solutions.

5.1.2 NO_3-contents in spinach

Without addition of nitrogen the NO_3-contents were low, 10-20 mmol per kg DM, except for experiment 7, where a content of 290 mmol per kg DM was found (Table 12, part e). With nitrate-nitrogen applied, the NO_3-contents increased; at the highest level, the contents were 500-1100 mmol per kg DM. Only in experiments 1B (first harvest) and 5, the contents at the highest N-levels were lower than 350 mmol per kg DM probably because of weather conditions. As was mentioned before (section 5.1.1), in 1977 a relatively dry period preceded the first harvest in experiment 1, probably resulting in a reduction in nitrate uptake, at least for the highest N-levels. In 1979, the weather during the growth period in Lelystad was relatively wet (section 3.3) and part of the NO_3-N might have been lost due to leaching.

In experiment 1B, no significant differences were found between NO_3-contents in spinach harvested at 37 and 44 days after emergence and neither any differences between NO_3-contents in spinach harvested in the 'morning' or in the 'evening' of a day with a radiation of 7 MJ per m^2.

In experiment 3, the NO_3-contents in spinach to which 150 or 200 kg NO_3-N per ha had been applied and which was harvested 30 days after emergence were lower than those in spinach harvested earlier. This effect may, however, have been caused by varietal differences.

In experiment 2B with 0-100 kg NO_3-N per ha applied, no differences in NO_3-contents between 'morning' and 'evening' harvests were found, but with 150 and 200 kg per ha applied, the NO_3-contents were about 100 mmol per kg DM lower in the evening. Radiation was about 24 MJ per m^2 on that day and the dry-matter contents increased to 8.4% at these nitrate levels. Obviously, a 'morning-evening' effect can be expected only on bright days with high ir-radiation and high NO_3-levels in the plants.

5.1.3 N-contents and total N in the aerial parts of spinach

Nitrate applications caused N(total)-contents to increase even more than NO_3-contents (Table 12, part f). With 200 or 225 kg NO_3-N per ha applied, increases of 900-2200 mmol per kg DM over values obtained in the zero-treat-ments were found, with relatively high N-contents encountered in experiments 3 and 7. NO_3-N accounted for about 0.5 to 27% of the N-contents (Table 12, part h), with the highest values (>20%) found in experiment 7.

Without added nitrogen, total N in the aerial parts of spinach was 2.0-4.0 g per m^2 (Table 12, part g) with an exceptionally high level of 11.3 g per m^2 in experiment 7. Total N in the aerial parts increased with the amounts of NO_3-N applied, the highest values always being found at the highest dressing. The amounts of NO_3 in the aerial parts increased with increasing quantities of N applied and were 0-2.5 g N per m^2 in Duiven and Lelystad and 0-4.5 g N per m^2 in Helden. The higher amounts in Helden are probably to be ascrib-ed to the presence of a sandy soil type with high levels of available N.

5.1.4 Available N in the soil profile

The quantity of available N present in a soil was estimated from the quan-tities of NO_3-N found to be present in the profile, augmented with the quan-tities of NH_4-N as determined in the soils of the 1979 experiments. In other years, the quantities of NH_4-N present were estimated from the results ob-tained in 1979. This procedure was followed for the so-called fallow plots as well as for the plots with spinach (Figure 5.2). The amounts of available ni-trogen during the growth period in the 0-60 cm layer of fallow plots without nitrogen applied are shown in Figure 5.1 (the choice of 0-60 cm depth will be discussed at the end of this chapter).

In Duiven, 41, 35 and 58 kg NO_3-N per ha were found before sowing in 1977, 1978 and 1979, respectively. NH_4-N was estimated at 14 kg per ha for all years (growth periods). As at the moments of final harvest, quantities of 85-90 kg (NO_3+NH_4)-N per ha were found to be present, the net mineraliza-tion per ha during the growth periods in Duiven could be estimated at 20-40 kg N.

Table 12. Effects of variations in nitrate dressing on yields (fresh and dry weight), dry-matter, NO_3^- and N-contents, total N, and the ratio (NO_3-N : N-total) × 100 in the aerial parts of spinach grown in the field experiments.

Experiment	1B	1B	1B	2B	2B	3	3	3	3	4	5	6	7
Variety	Estivato	Estivato	Estivato	Estivato	Indian Summer	Summic	Medania	Symfonie	731037	Spartan	Estivato	Estivato	Spartan
Days after emergence	31	37	44	35	24	26	30	30	30	39	39	45	45

NO_3-N kg per ha

a. Fresh yield (kg per m²)

NO_3-N	1B Estivato 31	1B Estivato 37	1B Estivato 44	2B Estivato 35	2B Indian Summer 24	Summic 26	Medania 30	Symfonie 30	731037 30	Spartan 39	Estivato 39	Estivato 45	Spartan 45
0	0.63	1.13	1.80	1.17	1.60	1.47	1.37	1.37	1.63	1.20	0.99	1.55	4.74
50	–	–	–	2.42	2.90	2.77	2.87	2.76	2.91	2.81	2.63	4.19	5.59
75	1.33	2.26	5.02	–	–	–	–	–	–	–	–	–	–
100	–	–	–	2.82	3.87	3.38	3.00	3.19	3.31	3.70	2.59	5.77	5.83
150	1.89	3.47	6.03	2.95	3.77	3.31	3.14	3.40	3.62	3.93	4.02	6.97	6.05
200	–	–	–	3.26	3.90	3.22	3.60	3.36	3.82	3.42	4.31	6.98	6.12
225	1.30	3.29	6.01	–	–	–	–	–	–	–	–	–	–

b. Fresh yield (% of yield with 150 kg N per ha)

NO_3-N	1B Estivato 31	1B Estivato 37	1B Estivato 44	2B Estivato 35	2B Indian Summer 24	Summic 26	Medania 30	Symfonie 30	731037 30	Spartan 39	Estivato 39	Estivato 45	Spartan 45
0	33	32	30	40	42	44	44	40	45	31	25	22	78
50	–	–	–	82	77	84	91	81	80	72	66	60	92
75	70	65	83	–	–	–	–	–	–	–	–	–	–
100	–	–	–	96	103	102	96	94	91	94	65	83	96
150	100	100	100	100	100	100	100	100	100	100	100	100	100
200	–	–	–	111	103	97	115	99	106	87	107	100	101
225	69	95	100	–	–	–	–	–	–	–	–	–	–

c. Dry-matter content (%)

NO_3-N	1B Estivato 31	1B Estivato 37	1B Estivato 44	2B Estivato 35	2B Indian Summer 24	Summic 26	Medania 30	Symfonie 30	731037 30	Spartan 39	Estivato 39	Estivato 45	Spartan 45
0	10.2	9.6	7.1	9.2	5.1	5.3	7.7	7.9	7.6	10.2	10.3	7.4	5.6
50	–	–	–	8.2	4.7	4.7	6.0	6.5	6.0	6.8	8.4	6.2	5.3
75	9.3	8.0	5.4	–	–	–	–	–	–	–	–	–	–
100	–	–	–	7.8	4.4	4.7	5.9	6.1	5.8	6.1	8.1	5.3	5.2
150	9.2	7.3	5.3	7.5	4.5	4.5	5.6	5.8	5.5	5.7	7.1	5.0	5.0
200	–	–	–	7.4	4.3	4.6	5.5	5.7	5.4	5.8	6.9	4.8	5.1
225	9.0	6.6	4.9	–	–	–	–	–	–	–	–	–	–

d. Dry-matter yield (g per m²)

NO_3-N	1B Estivato 31	1B Estivato 37	1B Estivato 44	2B Estivato 35	2B Indian Summer 24	Summic 26	Medania 30	Symfonie 30	731037 30	Spartan 39	Estivato 39	Estivato 45	Spartan 45
0	65	108	129	108	83	78	105	108	123	119	101	114	265
50	–	–	–	198	135	130	171	179	172	191	219	258	297
75	125	180	272	–	–	–	–	–	–	–	–	–	–
100	–	–	–	220	169	158	176	195	201	226	209	302	305

f. N (total) (m mol per kg DM)

	a	b								
0 / 50 / 75 / 100 / 150 / 200 / 225	285	609	896	1029	1069	3034	3460	3801	3920	3966
	14	41	211	583	791	1484	2201	2796	3516	3770
	6	29	60	212	348	1484	2015	2259	2833	3123
	14	100	354	619	646	1570	2575	3232	3678	3784
	8	128	246	483	554	2158	2859	2990	3762	3702
	9	44	270	524	521	2219	2512	3271	3788	3732
	7	119	322	324	559	2253	2888	3315	3425	3802
	36	171	363	699	751	2711	3293	3454	3861	3992
	27	122	326	641	770	2563	3003	3385	3676	4016
	15	65	170	443	595	1874	2270	2570	3205	3388
	15	117	356	608	1866	2845	3317	3709		
	13	78	342	620	1795	2416	3089	3714		
	13	113	238	280	2097	2746	2891	3067		

g. Total N in aerial parts (g per m²)

	a	b				
0 / 50 / 75 / 100 / 150 / 200 / 225	285/609	11.3	14.4	16.2	16.7	17.3
	14/41	2.4	8.0	11.8	17.2	17.7
	6/29	2.1	6.2	6.6	11.3	12.8
	14/100	2.7	6.9	10.2	11.6	10.6
	8/128	3.7	6.9	8.4	10.4	10.7
	9/44	3.3	6.3	8.9	10.4	10.0
	7/119	3.3	6.9	8.2	8.5	10.6
	36/171	3.0	6.0	7.6	8.0	8.2
	27/122	2.9	5.7	8.0	8.6	9.5
	15/65	2.8	6.3	7.9	9.9	11.6
	15/117	3.3	10.9	14.6	15.2	
	13/78	2.7	6.1	10.9	11.3	
	13/113	1.8	4.9	6.7	5.0	

h. (NO₃-N: N-total) x 100

0 / 50 / 75 / 100 / 150 / 200 / 225	9.4	17.6	23.6	26.2	27.0
	0.9	1.9	7.6	16.6	21.0
	0.4	1.4	2.7	7.5	11.1
	0.9	3.9	11.0	16.8	17.1
	0.4	4.5	8.2	12.8	15.0
	0.4	1.8	8.3	13.8	14.0
	0.3	4.1	9.7	9.5	14.7
	1.3	5.2	10.5	18.1	18.8
	1.1	4.1	9.6	17.4	19.2
	0.8	2.9	6.6	13.8	17.6
	0.8	4.1	10.7	16.4	
	0.7	3.2	11.1	16.7	
	0.6	4.1	8.2	9.1	

[59]

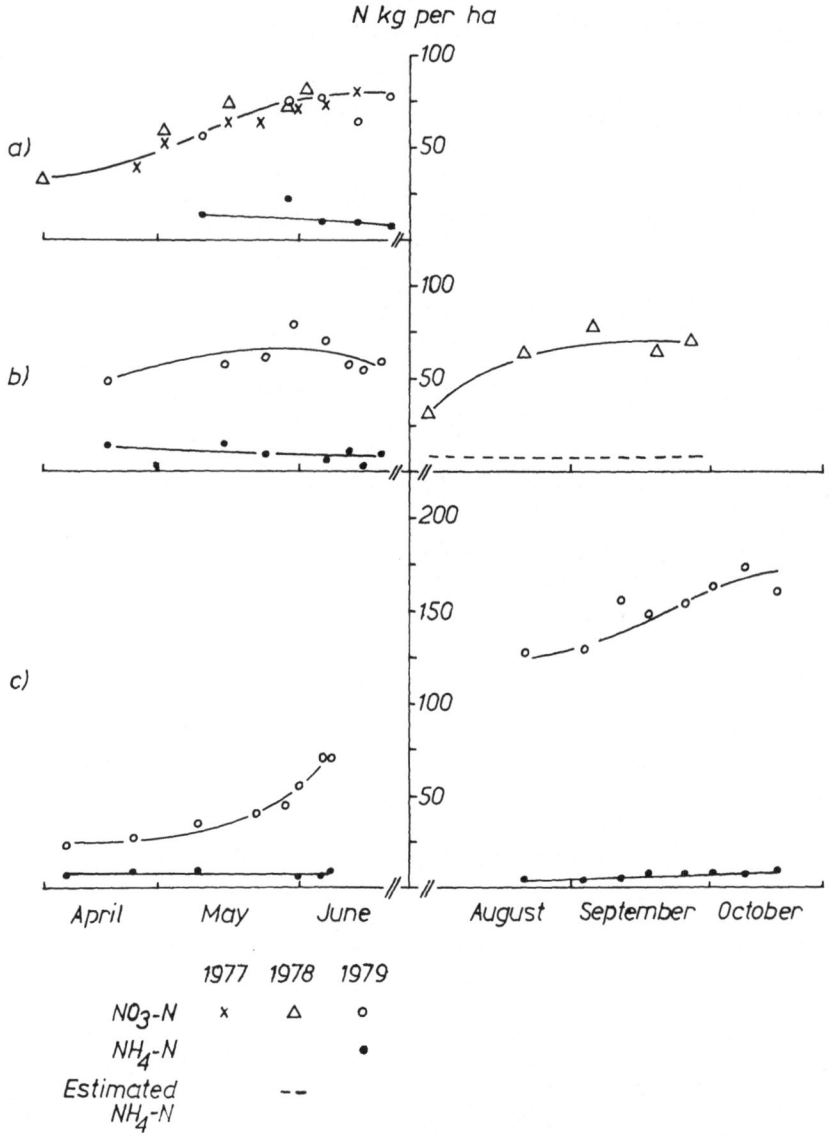

Figure 5.1 Available soil nitrogen (NO3-N and NH4-N) in the 0-60 cm layer of plots without plants and without nitrogen added.
a) Duiven b) Lelystad c) Helden.

In Lelystad in 1978, 32 and 69 kg NO_3-N per ha were found before sowing and at the point of final harvest, respectively, with the amount of NH_4-N estimated at 8 kg per ha. Hence, the net mineralisation must have amounted to about 35 kg N per ha. As in 1979 the amounts of NO_3-N and NH_4-N found to be present in Lelystad before sowing were 50 and 15 kg per ha, respectively, and at the final harvest these quantities were 61 and 8 kg per ha, it can be concluded that no net mineralisation and nitrification had taken place. Due to heavy rainfall, the soil conditions (temperature, moisture) were favourable for leaching and unfavourable for mineralisation.

The amounts of NO_3-N and NH_4-N in Helden in the spring of 1979 were 22 and 10 kg per ha, respectively, before sowing and 71 and 7 kg per ha, respectively, at the moment of final harvest. In Helden in the autumn of 1979, the amounts of NO_3-N and NH_4-N were 127 and 5 kg per ha, respectively, before sowing and 160 and 8 kg per ha, respectively, at final harvest time. In the Helden experiments, the net mineralisation was 30-50 kg N per ha based on spring quantities of 32 and 78 kg and autumn quantities of 132 and 168 kg (NO_3+NH_4)-N per ha being present before sowing and at final harvest, respectively. The high amounts present in Helden soil in the autumn of 1979 were due to past applications of manure up to 100 tons per ha per year, and to residual N resulting from a heavy dressing of fertilizer-N to a broccoli crop in 1979. High amounts of residual available N after vegetable cropping were also found by Scharpf [1978] and Böhmer [1980].

As can be seen from Figure 5.2, nitrogen taken up by the crop was withdrawn mainly from the 0-20 and 0-40 cm layers with minor amounts withdrawn from the 40-60 and 60-80 cm layers. Therefore, the top 60 cm of a profile were assumed to represent the soil body from which most of the nitrogen is taken up. The close correlation between the amounts of N in the plants and the quantities of NO_3-N that have disappeared from the 0-60 cm layers, as shown in Figure 5.3, support this assumption. The estimates of quantities of NO_3-N withdrawn from the top 60 cm of soil were calculated with the use of the following equation:

$$\frac{\Delta NO_3\text{-N}}{(0\text{-}6\ cm)} = \frac{NO_3\text{-N}(6\text{-}60\ cm)}{\text{on harvested plots}} - \frac{NO_3\text{-N}(0\text{-}60\ cm)}{\text{on fallow plots without N}} + \frac{NO_3\text{-N-dressing}}{\text{on havested plots}}$$

In her work with vegetables, including spinach, Böhmer [1980] used the 0-100 cm layer. Other findings from her work, however, indicate that spinach, especially in spring, withdraws nitrogen mainly from the 0-60 cm layer.

The earlier mentioned leaching of nitrogen from the soil profile in the 1979 Lelystad experiment can be observed in Figure 5.2.

5.1.5 Plant nutrients other than nitrogen

As can be seen from Figure 5.4, the contents of the individual cations and the sum of cations (in meq per kg DM) increased with increasing nitrate dressings. The contents of the individual inorganic anions decreased (Cl) or were not affected (H_2PO_4 and SO_4) with increasing nitrate dressings, but the sum of anions (NO_3 included) increased (Figure 5.4). As the increase in sum of cations was more pronounced than that in sum of anions, the difference (C-A) showed an increase. The ratio (C-A):org N varied in the experiments from 0.8 to 1.4 and decreased with increasing nitrate dressings.

[61]

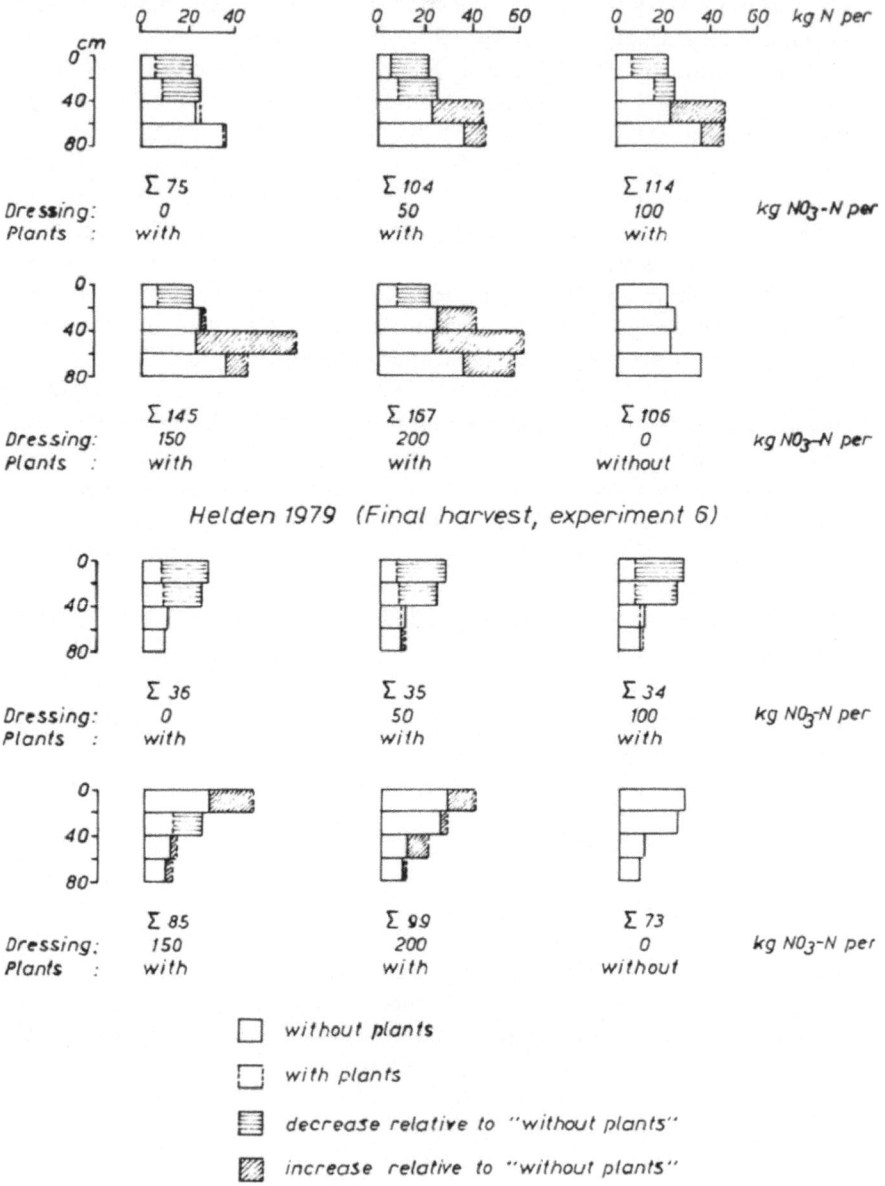

Figure 5.2 Available soil (NO_3+NH_4)-N in the soil profile of plots with spinach plants, having received varying quantities of NO_3-dressings, at the final harvests of experiments 5 and 6.

[62]

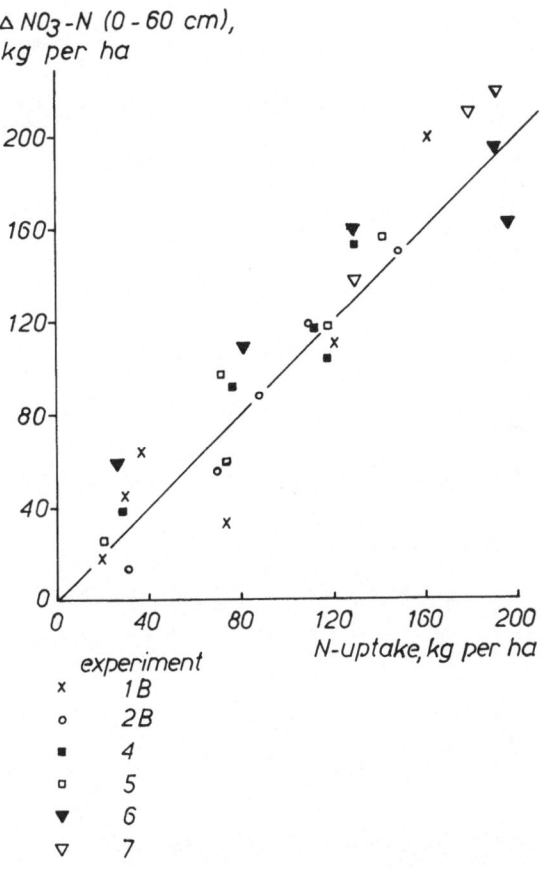

Figure 5.3 Relationship between NO_3-N having disappeared from the 0-60 cm soil layer (see text) and N-uptake by spinach in the field experiments.
(For the estimation of N-uptake it was assumed that total N in the aerial parts constitutes 90% of all nitrogen taken up).

Variations in nitrate dressing as well as in location affected the chemical composition. E.g., in Helden, the Ca-contents were about 100 mmol per kg DM lower than at the other locations, whereas the H_2PO_4-contents were about 100 mmol per kg DM higher. Furthermore, in Duiven the Na-contents were 500 mmol per kg DM higher than at the other locations and in Lelystad the Mg-contents were about 100 mmol per kg DM lower than at the other locations. Very remarkably, the K-contents in experiment 7 in general were about 500-700 mmol per kg DM higher than those in all other experiments (Figure 5.4). As a consequence of the increases in dry-matter yield with increasing nitrate dressings and of the above-mentioned effects on the contents of inorganic cations and anions and (C-A), the amounts of these ions (except for Cl) present per m^2 in the aerial parts of spinach also increased.

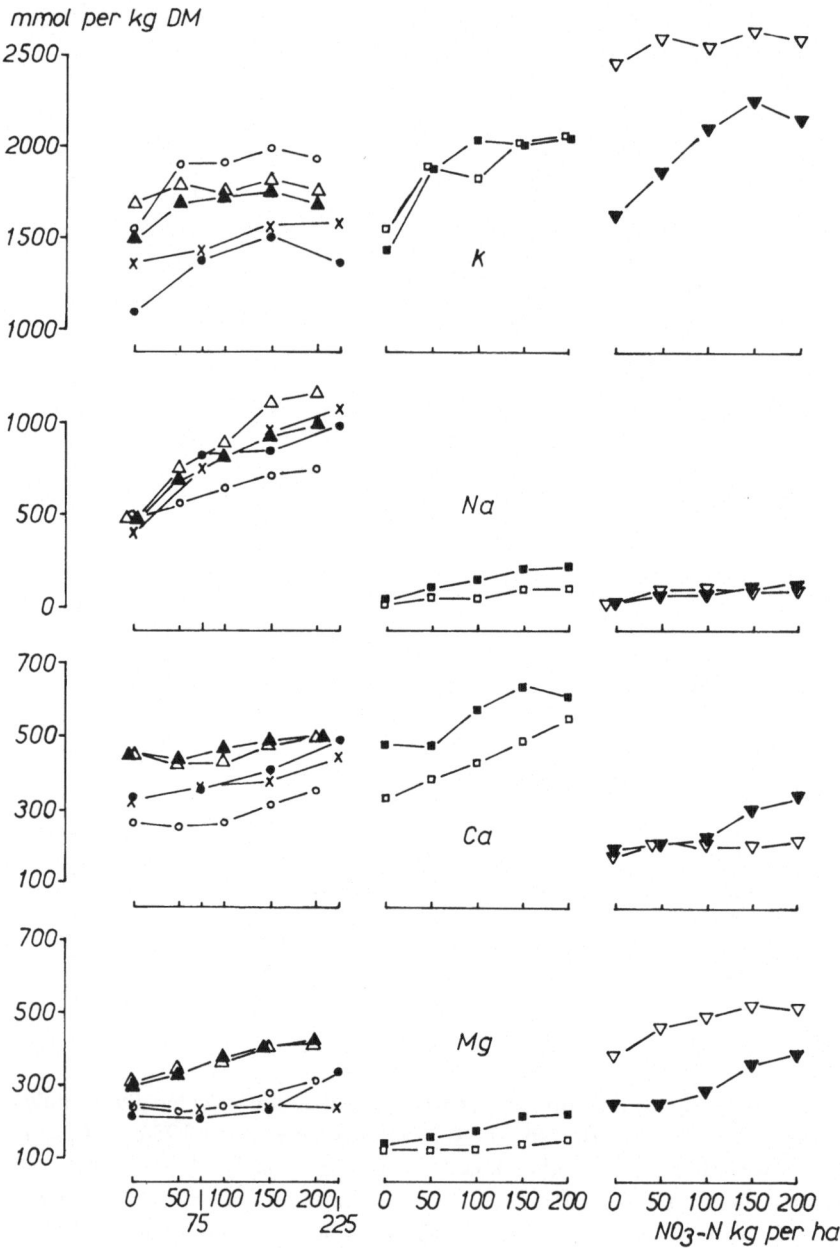

Figure 5.4 Effects of variations in nitrate dressings on the contents of K, Na, Ca, Mg, H_2PO_4, Cl, SO_4, C, A and (C-A) and the ratio (C-A):org N in spinach grown in the field experiments.

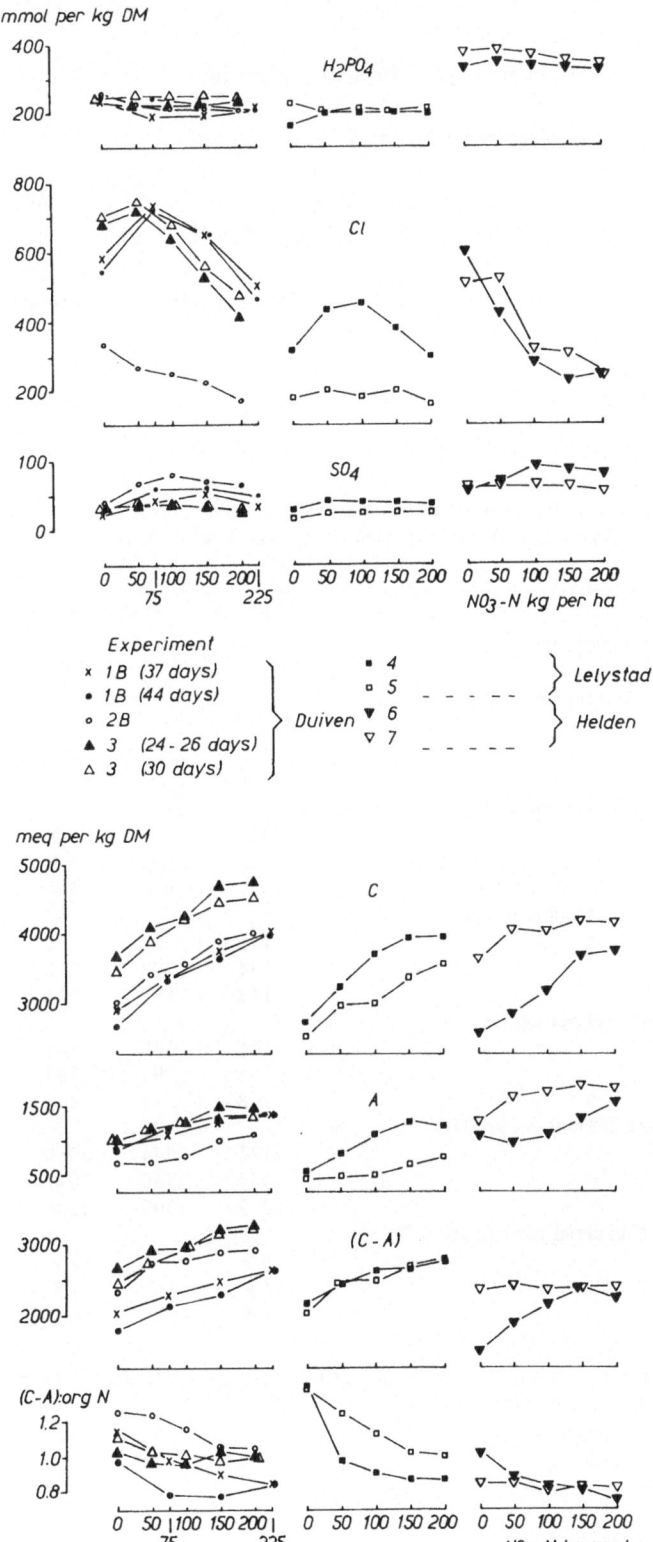

mmol per kg DM

H_2PO_4

Cl

SO_4

0   50 |100 150 200| 0   50 100 150 200   0   50 100 150 200
        75   225 NO_3-N kg per ha

Experiment
x   1B (37 days)   ■   4   } Lelystad
•   1B (44 days)   □   5   - - - -
○   2B   } Duiven ▼   6   } Helden
▲   3 (24 - 26 days)   ▽   7   - - - -
△   3 (30 days)

meq per kg DM

5000

4000

3000 C

1500

500   A

3000

2000 $(C - A)$

$(C-A)$:org N
1.2

1.0

0.8
0   50 |100 150 200| 0   50 100 150 200   0   50 100 150 200
        75   225 NO_3-N kg per ha

[65]

5.2 The effects of variations in timing of nitrogen applications

5.2.1 Yields and dry-matter contents of spinach

In experiment 1A (Table 13, parts a and c) yields were highest when all nitrogen was applied before sowing and decreased by 15-20% with increasing number of N-applications. With split applications, a tendency to higher dry-matter contents can be observed, especially at 31 days after emergence (Table 13, part b).

In experiment 2A (Table 14), additional dressings of NO_3-N and/or NH_4-N hardly affected yield.

Table 13. *Effects of variations in timing of nitrogen application and in plant age on yields (fresh and dry weight), dry-matter, NO_3- and N-contents and total N in aerial parts of spinach (Experiment 1A).*

	Harvest (days after emergence)	31	37	44
Treatment				
	a. Fresh yield (kg per m^2)			
D1[b])		1.61[a])	2.81	4.80
D3		1.51	2.45	4.87
D6		1.43	2.28	4.57
	b. Dry-matter content (%)			
D1		9.3	8.3	5.5
D3		9.6	8.4	5.3
D6		10.2	8.4	5.4
	c. Dry-matter yield (g per m^2)			
D1		148	227	260
D3		144	202	258
D6		142	191	244
	d. NO_3 (m mol per kg DM)			
D1		144	168	201
D3		58	94	187
D6		68	74	205
	e. N (total) (m mol per kg DM)			
D1		2724	2544	3060
D3		2430	2360	3097
D6		2259	2306	3085
	f. Total N in aerial parts (g per m^2)			
D1		5.7	8.2	11.2
D3		4.8	6.7	11.3
D6		4.5	6.1	10.5

[a]) Means of all N-forms and combinations with and without nitrapyrin. N: 150 kg per ha.
[b]) D1: N in one basal application before sowing.
    D3: N split in three applications.
    D6: N split in six applications.

Table 14. *Effects of variations in additional dressings of NO_3-N and/or NH_4-N on yields (fresh and dry weight), dry-matter, NO_3- and N-contents and total N in aerial parts of spinach with a basal dressing of 100 kg of NO_3-N per ha before sowing (Experiment 2A).*

Additional dressings[a]						
NO_3-N (kg per ha)	0	50	100			50
NH_4-N (kg per ha)				50	100	
Fresh yield (kg per m^2)	2.54	2.54	2.76	2.77	2.56	2.63
Dry-matter content (%)	8.3	7.9	8.1	7.9	8.0	8.0
Dry-matter yield (g per m^2)	211	201	222	220	206	211
NO_3 (m mol per kg DM)	259	371	460	300	335	336
N (total) (m mol per kg DM)	2912	3106	3330	3019	3140	3153
Total N in aerial parts (g per m^2)	8.6	8.8	10.4	9.3	9.0	9.3

[a]) Applied 13 days after emergence.

5.2.2 NO_3- and N-contents and total N in aerial parts of spinach

In comparison with all N applied as a basal dressing, a partitioning of the N in split dressings caused the NO_3- and N-contents and total quantities of N in the aerial parts of spinach in experiment 1A to be notably lower at 31 and 37 days after emergence but not at the final harvest (Table 13 parts d, c and f). From 31 to 44 days after emergence, the NO_3-contents increased.

In experiment 2A, additional dressings of NO_3-N gave higher NO_3- and N-contents and higher total N values in the aerial parts of spinach than did comparable dressings of NH_4-N or (NH_4+NO_3)-N (Table 14). Without an additional N-dressing, so only within a basal dressing of 100 kg NO_3-N per ha, the NO_3- and N-contents and total N in the aerial parts in experiment 2A (Table 14) were lower than in experiment 2B (Table 12, parts e, f and g) with the same dressing. These differences cannot be explained.

5.3 The effects of variations in NO_3:NH_4-ratio

5.3.1 Yields and dry-matter contents of spinach

In experiments 1A, 4 and 6, yields (fresh and dry weight) decreased with decreasing NO_3-N:NH_4-N-ratios at N-dressings of 100 and 150 kg per ha. The decreases were most pronounced in the treatments with nitrapyrin (Table 15, parts a and c). Compared with spinach yields obtained with NO_3-N only and without nitrapyrin, a gradual replacement of NO_3-N by NH_4-N caused fresh yields to decrease with 52, 23 and 37% and dry-matter yields with 41, 11 and 27%. The yield decreases were less than 20% with NO_3-N:NH_4-N-ratios higher than one.

The dry-matter contents increased with decreasing NO_3-N:NH_4-N-ratios in the experiments mentioned (Table 15, part b) and this explains why the decrease in dry weight is less than that in fresh weight.

None of these effects were found in experiment 5, (150 kg N applied) and experiment 7 (100 and 150 kg N applied).

5.3.2 NO₃-contents in spinach

The NO_3-contents in spinach generally decreased with decreasing NO_3-N: NH_4-N-ratios, except in experiments 5 and 7 (150 kg N per ha). In experiment 5, the NO_3-contents even increased when a part of the NO_3-N was replaced by NH_4-N (Table 15, part d). With nitrapyrin added, the decrease in NO_3-contents was substantial in experiments 4 and 6, but in experiments 1A and 7, no effects of nitrapyrin were found. In case of positive effects of nitrapyrin, compared with 100% NO_3-N, a 1:1 ratio of NO_3-N and NH_4-N caused 30 to 50% decreases in nitrate contents.

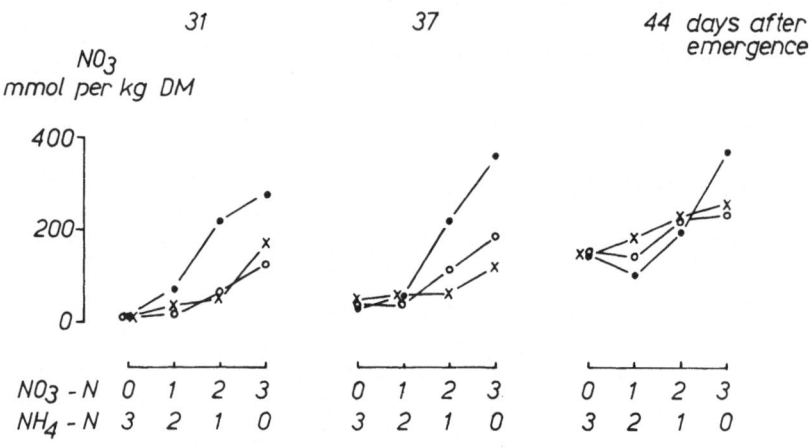

Figure 5.5 Effects of variations in timing of N-application, in NO3:NH4-ratio and in age of plants on the NO3-contents in spinach.
(Data presented are means of values obtained with and without nitrapyrin, N-dressing is 150 kg per ha)(Experiment 1A).

In experiment 1A, an interaction between timing of N-application and NO_3-N: NH_4-N-ratio was found (Figure 5.5), especially for the data obtained 31 and 37 days after emergence. At these two harvests, the NO_3-contents for the low NO_3-N:NH_4-N-ratios were about equal, independent of timing of N-application, whereas for the higher NO_3-N:NH_4-N-ratios higher NO_3-contents were found, at least when all N was applied as a basal dressing. Nitrapyrin did not influence this effect.

The interaction probably must be attributed to variations in the mobility of the different N-forms. A rapid fixation of NH_4, applied either as a basal dressing or as a late dressing, by the clay loam must have taken place and at least up to 37 days after emergence, this initial fixation must have had a retarding effect on nitrification and subsequent translocation of NO_3-N to the spinach roots (NH_4 was not worked in in this experiment). The appreciable decrease in NO_3-contents (and in yields) with a lower NO_3-N:NH_4N-ratio, independently of nitrapyrin addition, supports the view that NH_4 was strongly adsorbed onto the clay minerals.

The second half of May 1977 was characterized by dry weather. The transport to the root zone of NO_3-N applied in that period (i.e. up to 37 days after emergence) must have been less than that of NO_3-N applied as a basal dressing. As a result of sufficient rainfall between 37 and 44 days after emergence, the differences in NO_3-contents more or less disappeared later on. Leaching of NO_3-N (e.g. in experiment 5 with a wet growth period) had the same effects on spinach (i.e. reduction of the NO_3-contents) as had a lower NO_3-N:NH_4-N-ratio combined with nitrapyrin. Without addition of nitrapyrin, a portion of the NH_4-N must have been adsorbed onto the clay minerals, which prevented it from leaching but caused it to be nitrified belatedly thus accounting for the higher NO_3-contents.

5.3.3. N-contents and total N in the aerial parts of spinach

In general the N-contents in the aerial parts of spinach were affected by the NO_3-N:NH_4-N-ratios in the same way as were the NO_3-contents. (Table 15, part e). The decreases in N-content amounted to about 400 mmol per kg DM when the share of NO_3-N is the total N applied dropped from 100% to 50% (experiments 1A, and 6, with 100 kg N applied).

Total N in the aerial parts also decreased with a decrease in NO_3-N: NH_4-N-ratio (except in experiment 5), especially when nitrapyrin was added (Table 15, part f).

In experiment 1A, at 37 days after emergence the reduction in total N was higher than in any other experiment, i.e. up to 5.9 g per m^2.

5.3.4 The ratio (C-A):org N in the aerial parts

The effects of variations in NO_3-N:NH_4-N-ratio on the (C-A):org N-value (meq per mmol) in the aerial parts of spinach are demonstrated in Figure 5.6. No such effects were found in experiment 7, neither with nor without nitrapyrin, in experiment 1A at 31 and 37 days after emergence and in experiments 4 and 5 for the treatments without nitrapyrin. With increases in the NO_3-N:NH_4-N-ratio, the (C-A):org N-value however increased in experiments 4 and 5 in the treatments with nitrapyrin, whereas in experiments 1A (final harvest) and 6 this occurred with and without nitrapyrin addition.

In the latter two experiments, the (C-A):org N-value was higher without nitrapyrin (Figure 5.6, c and f).

[69]

Table 15. Effects of variations in quantity of N-dressing, in $NO_3 : NH_4$ - ratio and in plant age on yields (fresh and dry weight), dry-matter, NO_3- and N-contents and total N in the aerial parts of spinach grown in the field experiments

N-dressing (kg per ha)	150														100			
Experiment	1 (Aa))						4		5		6		7		6		7	
Days after emergence	31		37		44		39		39		45		45		45		45	
NO_3-N : NH_4-N / Nitrapyrin b)	0	1	0	1	0	1	0	2	0	2	0	2	0	2	0	2	0	2
a. Fresh yield (kg per m²)																		
NH4 only	1.08	1.00	1.54	1.67	3.99	3.62	3.55	3.04			4.45	4.42	5.92	5.76	4.67	4.12	5.77	5.82
												5.57		6.04		4.61		6.00
1 : 3	1.53	1.42	2.40	2.02	4.37	4.45		4.09										
1 : 2								4.35	4.33		5.56	5.16	6.16	5.36	5.18	5.36		
1 : 1	1.64	1.86	2.91	3.03	4.83	5.43		4.45			5.95	5.56	5.79	6.16	5.18	5.36	5.97	6.26
NO3 only	1.84	1.78	3.19	3.37	5.76	5.50	3.93	3.54	4.01	4.14	6.97	6.48	6.05	6.13	5.77	5.45	5.83	5.94
																		6.26
b. Dry-matter content (%)																		
NH4 only	11.3	10.5	9.1	9.0	5.6	5.6	5.8	6.6	6.7	6.7	5.2	5.8	5.0	5.0	6.0	6.0	5.2	5.1
												5.5		5.0		6.1		5.2
1 : 3	9.9	9.8	8.5	8.7	5.4	5.7		7.0	6.9	7.0	5.3	5.4	5.2	5.1	6.1	6.0		
1 : 2							6.3	7.0										
1 : 1	9.2	8.9	8.3	7.8	5.2	5.4	5.9	6.9							5.3	5.9	5.2	5.3
							6.2											
NO3 only	9.0	9.2	7.5	7.9	5.2	5.0	6.0	7.0	7.1	7.0	5.0	5.0	5.0	5.2	5.3	5.6	5.2	5.3
c. Dry-matter yield (g per m²)																		
NH4 only	121	105	141	150	223	204	204	201	285	280	308	296	290	278	246	301	300	
												302	304		279		314
1 : 3	151	139	204	176	237	251		205	292	283	314	301	299	313	314	320	312	321
1 : 2							205	302										
							204	306										
1 : 1	151	166	242	236	253	291	206		301									
2 : 1																		

	NH$_4$ only	1:3	1:2	1:1	2:1	3:1	NO$_3$ only
	19	48	125	178			
	13	35	101	206			
	47	55	101	228			
	37	47	167	213			
	197	149	248	261			
	97	140	180	311			
	502	619					
	161	253 / 291 / 377	586				
	378	237	212				
	198	199 / 174 / 210	185				
	401	469	583				
	143 / 243	386	407 / 555				
	1037	884	1028				
	910 / 860	1001	940 / 954				
	108	114	211				
	52 / 73	94	192 / 244				
	701	684	896				
	652 / 733	724	785 / 825				

e. N (total) (m mol per kg DM)

	NH$_4$ only	1:3	1:2	1:1	2:1	3:1	NO$_3$ only
	1979	2367	2656	2826			
	2160	2329	2649	2808			
	2090	2196	2371	2777			
	2169	2230	2616	2763			
	3106	3045	3119	3087			
	3079	2905	2924	3381			
	3606	3678					
	3338	3556 / 3562 / 3456	3666				
	3024	2820	2832				
	3154	2991 / 2714 / 2812	2608				
	3732	3800	3516				
	3283 / 3410	3712	3506 / 3618				
	3906	3821	3921				
	3831 / 3773	3867	3799 / 3909				
	2677	2734	2796				
	2829 / 2629	2789	3006 / 3169				
	3553	3592	3801				
	3565 / 3696	3660	3616 / 3641				

f. Total N in aerial parts (g per m^2)

	NH$_4$ only	1:3	1:2	1:1	2:1	3:1	NO$_3$ only
	3.2	4.7	5.6	6.4			
	3.1	4.5	6.1	6.3			
	4.1	6.3	7.9	9.3			
	4.5	5.4	8.6	10.4			
	10.0	10.2	11.0	12.8			
	8.7	10.1	11.8	13.1			
	10.3	11.6					
	9.4	10.2 / 10.2 / 10.0	10.9				
	12.3	11.7	11.3				
	12.4	12.0 / 11.5 / 12.0	10.6				
	16.0	16.7	17.0				
	11.8 / 14.4	15.7	15.7 / 16.1				
	16.0	16.2	16.7				
	15.5 / 16.1	16.9	16.6 / 17.3				
	10.4	12.0	11.8				
	9.7 / 10.3	12.5	12.9 / 13.5				
	15.0	15.7	16.2				
	15.0 / 16.2	16.4	15.8 / 16.6				

a) Averaged over all application dates.
b) Percentage by weight of N applied.

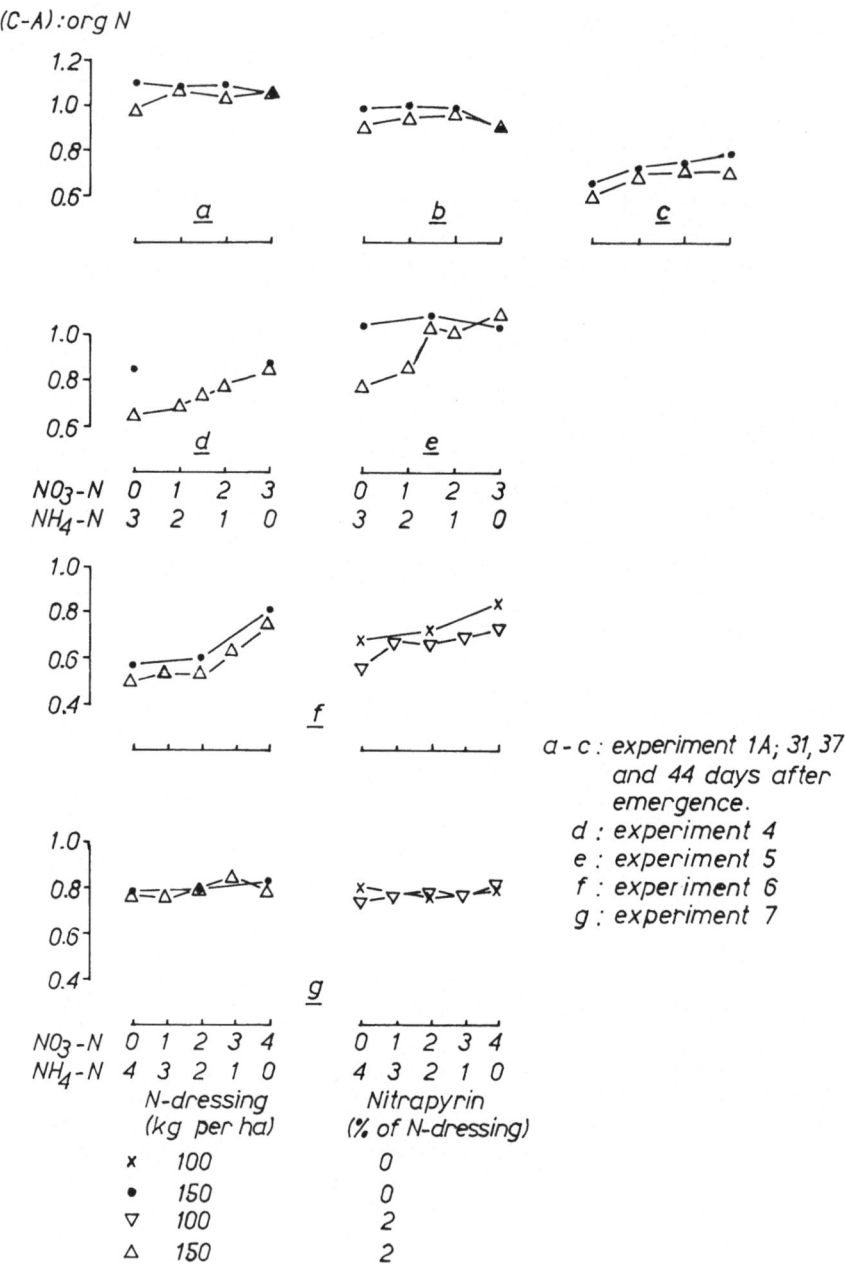

Figure 5.6 Effects of variation in NO_3-N:NH_4-N-ratio on the (C-A):org N-value in the aerial parts of spinach grown in the field experiments.

5.4 Experiments on farmers' fields

In spring 1980 simple experiments were carried out on farmers' fields. The fields were located at 15 different sites in the Netherlands and ranged from 30 to 151 kg per ha in available N present before sowing in the 0-60 cm soil layer. Each field was split in two units, one receiving a 'high' N-dressing and one a 'low' N-dressing, whereby the dressings varied from field to field. The 'high' N-dressings (high-N) were in accordance with farmers' practices, whereas the 'low' N-dressings (low-N) were based on the results of the field experiments carried out in 1977-1979. The latter dressings were chosen in such a way that the amount of plant-available N (= N-dressings plus the amount of available N in the soil before sowing) was about 150 kg per ha.

In Table 16 details on the experiments are presented. At harvest time, per field unit 4 samples of spinach were taken, each comprising $2 m^2$. The spinach plants were cut at 2-3 cm above the soil surface according to practice. The samples were weighed, dried and analyzed for NO_3 as described in section 3.5.

5.4.1 Yields and dry-matter contents of spinach

With a 'low' N-dressing, yields were on the average 17% lower than with a 'high' N-dressing (Table 16). Mainly on the sandy soils, these yield differences were appreciable, which probably must be attributed to leaching of N from these soils. For these sandy soils leaching will have a more pronounced effect in the low-N than in the high-N-treatments.

On all fields, the dry-matter content was higher with a 'low' N-dressing than with a 'high' N-dressing, on the average the difference being 10%. The low dry-matter contents on field nr. 12 are probably a reflection of the young age of this crop at harvest.

5.4.2 NO_3-contents in spinach

On all fields the NO_3-content was lower with a 'low' N-dressing than with a 'high' N-dressing, the average difference being 69%. The very low NO_3-contents with a 'low' N-dressing on fields nr. 2, 4 and 5 justify the assumption that here leaching of N must have occurred. The relatively high NO_3-content with a 'low' N-dressing on field nr. 12 is to be ascribed on the one hand to the young age of the crop and on the other hand possibly to a high mineralisation rate of organic N, applied in the past as manure.

More details on the results of the experiments on farmers' fields are presented by Titulaer [1980].

Table 16. Effects of variations in quantity of N-dressing on yield (fresh weight), dry-matter and NO$_3$-contents of spinach grown on farmers' fields showing variations in soil type, sowing date, seed amount, spinach variety, harvest date and available soil N before sowing.

Farmer	Soil type	Sowing date	Seed amount (kg per ha)	Variety	Harvest date	Available soil N (kg per ha)	N-dressing (kg per ha)		Yield (kg FW per m^2)		Dry-matter content (%)		NO$_3$-content (mmol per kg DM)	
							Low-N	High-N	Low-N	High-N	Low-N	High-N	Low-N	High-N
1	sand	27 February	100	Indures	12 May	93	60	129	2.50	2.93	8.5	7.6	183	315
2	sand	28 February	100	Jovita	12 May	73	75	267	1.41	3.07	10.5	8.0	29	500
3	sand	26 March	80	Butterflay	21 May	37	96	264	2.66	3.86	8.0	7.5	154	664
4	sand	24 March	80	Spinoza	22 May	50	90	300	1.70	2.58	11.5	9.7	30	582
5	sand	5 March	100	Nowak	22 May	52	101	197	1.81	2.41	9.9	9.2	36	93
6	marine clay	25 March	70	Spinoza	26 May	42	104	247	1.99	1.98	8.9	8.0	86	249
7	river clay	24 March	50	Butterflay	27 May	151	0	65	1.99	2.66	9.3	8.9	30	108
8	river clay	10 April	65-70	Butterflay	27 May	124	52	210	3.03	3.75	7.3	6.2	240	840
9	river clay	8 April	50	Summic	28 May	46	104	182	2.80	3.15	7.3	7.0	198	855
10	river clay	8 April	65-70	Butterflay	29 May	116	36	298	2.66	2.83	6.7	6.6	155	484
11	river clay	10 April	50	Summic	2 June	46	104	156	2.61	3.15	8.2	7.7	134	239
12	marine clay	24 April	60	Symphonie	3 June	105	52	220	3.13	3.47	5.4	5.3	541	648
13	marine clay	14 April	50	Nowak	3 June	43	104	182	2.63	2.75	7.8	7.5	99	203
14	marine clay	12 April	70	Butterflay	4 June	30	120	266	2.69	3.00	7.9	6.9	134	347
15	marine clay	11 April	50	Nowak	11 June	36	124	186	3.30	3.14	7.4	7.3	281	398

6 Discussion

Many vegetables, including spinach, possess the unpleasant property of accumulating nitrate in excess of the quantity needed for an adequate protein synthesis. Since high nitrate intakes can be considered to be a hazard to human health, it is of utmost importance to investigate whether these vegetables can be grown in such a way that an acceptable production level can be maintained without the frequently occurring high nitrate contents.

Cultural measures taken to attain this goal can be divided into four categories. One includes measures aimed at making use of or manipulating environmental and soil conditions in such a way that nitrogen-fertilization levels needed to bring about optimal growth can be maintained with avoidance of unacceptably high NO_3-levels. A second comprises measures taken to make use of the finding that cultivars differ in their tendency to accumulate nitrate. The third category consists of measures aimed at circumventing the characteristics of these vegetables to accumulate nitrate by regulation of nutritional conditions other than those governed by nitrogen. In the fourth category, those measures are listed that are taken to regulate the quantity, the form, and the timing of application of nitrogen fertilizer in such a way that high yield levels are maintained without an ensuing accumulation of nitrate in the vegetable tissue.

Since the majority of experiments conducted in the present investigation can be placed in the fourth category, the former three categories will be taken together, and the results obtained will be discussed in one section, preceding the section in which the experiments dealing with N-dressings are covered.

6.1 Cultural measures, other than nitrogen dressings, aimed at minimizing NO_3-contents in spinach

When in water-culture experiments, at an advanced stage of growth nitrate was withheld from spinach plants over a 9-day period, it was found that the plants were able to utilize the previously stored nitrate for protein synthesis. During this period, the dry-matter yield remained more or less stable, but the fresh weight of the aerial parts declined by 25%. These results make it clear that even under circumstances enabling a grower to terminate the supply of nitrate to the plants, which under field conditions is usually not easily achieved, little benefit is to be expected from such an achievement. The results prove that most nitrate stored in the older leaves is labile enough to be utilized for amino-acid synthesis, but that such a belated consumption of stored nitrate reduces the turgescence of the leaves and thus leads to a reduction in weight of the produce.

In spinach plants, nitrate tends to accumulate in older leaves (Figure 4.7). It is experienced, however, that the nitrate content in the harvestable portion of the plant decreases with age. Especially in spring-grown spinach, extending the growth period may be helpful in reducing the nitrate content. Not only does such an extension result in a more mature crop, but due to gradually increasing light intensities it also raises the photosynthesis rate, whereby the incorporation of nitrates into proteins is enhanced. Obviously, any extension of the growth period of autumn-grown spinach may have the opposite effect, since as a result of declining light intensity the plant is increasingly facing difficulties in utilizing nitrates for protein synthesis (Figure 4.9).

In growth-chamber experiments, the effect of light intensity on NO_3-content in spinach leaves was very evident (Figure 4.11), but in the field experiments there was only one occasion (experiment 2B) at which as a result of very bright weather the nitrate content in spinach harvested in the evening was distinctly lower than in spinach harvested in the morning.

Likewise, growers will not often be in a position to make use of the finding that NO_3 tends to accumulate less in spinach leaves during cold than during warm days. Nevertheless, it may be useful for growers to keep in mind that the quality of spinach is enhanced when harvested on cold, bright days. The advantage arising from low temperature is likely to be associated with a reduction in respiration, thus allowing more nitrogen to react with photosynthates in forming proteins.

On theoretical grounds, there are several reasons for assuming that high soil-moisture contents will be helpful in reducing the NO_3-contents in spinach shortly before harvest. In the first place, at high soil-moisture levels, the concentration of NO_3 in the soil moisture will be reduced which may lead to a slowdown in the NO_3-absorption rate. Second, at high moisture levels, conditions may become favourable for denitrification causing the NO_3-availability to be reduced. Third, applying additional water by irrigation shortly before harvest may be helpful in flushing NO_3 into soil layers out of reach of the spinach roots.

In the field experiments, soil moisture was never used as an experimental variable, so that no statements can be made about the effect of variations in moisture content. In a pot experiment (Table 7), however, it was observed that a subnormal moisture level raised the NO_3-content in spinach, but that a higher than normal moisture level did not cause the NO_3-contents to decline. In addition, it must also be borne in mind that from a practical standpoint serious disadvantages may be attached to imposing high soil-moisture levels shortly before spinach is to be harvested.

The relatively high moisture-holding capacities of medium-textured in comparison with light-textured soils may also be responsible for the observation, made both in the present study (Figure 4.5, experiment 10) for spinach and in Germany for vegetables other than spinach [Geyer, 1978], that NO_3-

contents in vegetables are lower when grown on the former than on the latter soils. As a result of these higher moisture-holding capacities, NO_3-concentrations in the moisture of medium-textured soils can be expected to be comparatively low.

It was furthermore found that spinach grown on a neutral soil contained less NO_3 than when grown on an acid soil (Figure 4.10). It might be that the frequently experienced temporary immobilization of added fertilizer nitrogen, often attributed to bacterial action, is more complete in neutral soils with their relatively high bacterial population than in acid soils. Such a temporary immobilization might affect the N-nutrition of a rapidly growing, short-season crop, like spinach, more than it does the N-nutrition of most other crops with their longer growing seasons.

The higher NO_3-contents in spinach grown on acid soils in the present investigation could not be attributed to a limited availability of Mo (section 4.4.7). Mo is an element needed for the proper functioning of the enzyme nitrate reductase involved in the reduction of NO_3 preceding the incorporation of the nitrogen into amino-acid [Beevers & Hageman, 1969]. The absence of any effect of Mo-addition on NO_3-contents in spinach leaves of the present experiment, of course, does not exclude the possibility that lack of available Mo sometimes is responsible for high NO_3-levels in spinach grown on acid soils.

The well-known positive influence of K on the uptake of NO_3 was also experienced in this study. K-application stimulated N-uptake and yield of spinach, but total N-contents were decreased. As the NO_3-contents in general increased with rising K-applications (Table 5, Figure 4.14), there is a reason to assume that K has a suppressive effect on NO_3-reduction inside the plant.

The stimulatory effect of K on N-uptake can be suppressed when the K is applied in the form of KCl. More than the SO_4-ion, the Cl-ion is capable of competing with NO_3 for attachment to absorption sites. As a result, lower NO_3-contents were found with KCl than with K_2SO_4.

The impression obtained in the greenhouse experiment with different spinach varieties was that the observed variations in NO_3-content were primarily a reflection of differences in growth rate (Table 11). Under field conditions, such variations in NO_3-content were also observed (Table 12, part e, experiment 3), but here the variations could not completely be ascribed to varietal differences, as harvest dates were not identical. Judging from the experiences of Cools et al. [1980] who observed inconsistent results with spinach cultivars in greenhouse studies in the Netherlands, the importance of the observed differences in tendency to accumulate NO_3 is still questionable. Nevertheless, even when there is as yet little reason to postulate that currently used spinach cultivars exhibit genetically regulated differences in rate of NO_3-accumulation or rate of NO_3-reduction, from a practical standpoint it is important to pay attention to differences in growth characteristics which may indirectly lead to differences in NO_3-accumulation and in NO_3-content of the produce.

[77]

High NO_3-contents can also be avoided when petioles and older leaves are excluded from the harvested material (Figure 4.7). In a sense, the advantages to be gained from such selective harvesting, can be compared with those resulting from a postponement of the harvest date.

6.2 Cultural measures related to nitrogen, aimed at minimizing NO_3-contents in spinach

An obvious conclusion to be drawn from the observation that spinach tends to accumulate nitrate is, that attempts should be undertaken to accommodate the N-nutrition and the growth of spinach through the supply of NH_4 that is protected against nitrification. However, a complication immediately encountered in such attempts is the observation that spinach belongs to the group of plants responding very unfavorably to NH_4-nutrition.

The culture of spinach on sand with NH_4 as sole source of nitrogen turned out to be unsuccessful. When NH_4 was partially replaced by NO_3, the resulting lower NO_3-contents in spinach were accompanied by reductions in yield (Figures 4.2 and 4.3). The data obtained in experiment B2 show that a partial replacement of NO_3 by NH_4 is no guarantee for lower NO_3-contents, not even in a sand culture in which nitrification is likely to be absent.

It may be postulated that with a partial replacement of NO_3 by NH_4, on account of a lower energy investment the spinach plant prefers NH_4 to NO_3 for incorporation into protein resulting in an accumulation of NO_3 inside the plant, with dry-matter yield not being affected by the partial replacement.

When spinach is grown on soil, nitrification inhibitors are commonly needed to protect NH_4 from being nitrified. With the use of these nitrification inhibitors, the NO_3-contents in spinach leaves can be suppressed, but such a suppression is usually accompanied by a yield decline [Mills et al., 1976; Sommer & Mertz, 1974]. For spinach grown on sandy soils with a pH below 7 this decline is likely to be caused by a combination of two adverse effects brought about by $(NH_4)_2SO_4$-dressings: first, soil acidification, even in the absence of nitrification, resulting from the hydrolysis of $(NH_4)_2SO_4$ and the acidic uptake pattern brought about by the uptake of NH_4-N, and second, the poor response of spinach to NH_4-nutrition. The decrease in soil pH in the absence of nitrification could be clearly demonstrated in a soil-culture experiment (Table 8).

For spinach grown on other soils the situation may be different, i.e. the decrease in soil pH may be less pronounced and part of the NH_4-N may be adsorbed onto clay minerals. An extreme example of the latter was shown in a soil-culture experiment using a clay-loam soil. Here certain advantages seemed to be attached to a partial replacement of NO_3 by NH_4 (Table 9). Yields were not suppressed by such a replacement, whereas NO_3-contents in leaves declined when more than half of the N was applied as NH_4. Upon closer examination of the experimental conditions it turned out, however,

[78]

that the soil involved had a capacity to fix NH_4. Consequently, a partial replacement of NO_3 by NH_4 had the effect of a reduction in N applied, so that the results obtained were more a reflection of a decreased N-uptake than of a shift in NO_3:NH_4-uptake ratio. This finding is of particular interest when compared with the results of other soil-culture experiments in which it was also frequently observed that the NO_3-content in spinach leaves is most effectively reduced by diminishing the quantity of NO_3 applied (Figure 4.5).

From the soil-culture experiment with clay-loam soil as well as from the sand-culture experiments it could be concluded that in addition to the NO_3: NH_4-ratio, also the amount of N applied and some other growth factors, such as light, affect the NO_3-contents and yields of spinach. Obviously, with a high level of available N a partial replacement of NO_3-N by NH_4-N is not effective, since with such a replacement the NO_3-availability remains high enough to cause luxury consumption of NO_3. At low levels of available N, a partial replacement of NO_3-N by NH_4-N may reduce the absorption of NO_3 sufficiently to result in lower NO_3-contents and lower yields.

Under field conditions the level of N-availability is not determined only by the quantity of N present in soil before sowing (mainly NO_3-N, Figure 5.1) but is also affected by mineralization of soil organic N, by leaching of N, predominantly NO_3, by NH_3-volatilization, by denitrification, and by the effectiveness of added nitrapyrin.

From the results of the field experiments 5 (leaching of NO_3-N) and 7 (high levels of available N in soil before sowing), as presented in Table 15, it may be concluded that the addition of NO_3-N and/or NH_4-N in combination with nitrapyrin is not necessarily reducing the NO_3-contents in spinach. When in this respect a positive effect is found, it is usually accompanied by a reduction in yield, as was also observed by Bengtsson [1968].

The use of nitrapyrin under field conditions is accompanied by a number of uncertainties rendering the effect of nitrapyrin rather unpredictable. Volatilization [Briggs, 1975] and segregation of nitrapyrin and NH_4 [Rudert & Locascio, 1979] may occur in the field. Furthermore, as Kerkhoff & Slangen [1980] indicate in their review, soil texture, organic-matter content, pH, moisture content and temperature can influence the effectiveness of nitrapyrin in soil.

Without the use of a nitrification inhibitor, a partial or complete replacement of NO_3 by NH_4 can be expected to modify the uptake of NO_3 when this is formed gradually from NH_4 due to nitrification. It can be reasoned that in this way a situation in which the rate of NO_3-uptake exceeds the rate of NO_3-reduction, might be avoided. Experiments conducted to test the validity of this reasoning, however, proved unrewarding. In a soil-culture experiment the soil-pH decrease resulting from nitrification caused yield to decline leading to higher rather than lower NO_3-concentrations in the leaves (Table 8). In the field, however, without the use of a nitrification inhibitor, in general lower NO_3-contents but also lower yields were obtained with the partial or complete replacement of NO_3 by NH_4 (Table 15).

The favourable results obtained with slow-release N-fertilizers in a pot experiment (Figure 4.12) may have little practical value, as these fertilizers appeared to be primarily effective in preventing high salt concentrations in the early growth stage of the spinach plants, a situation often encountered in pot experiments but not so likely to arise under field conditions, except with excessively high NO_3-dressings (see Table 12). The promising results obtained with farmyard manure and pig-manure slurry in a pot experiment (Figure 4.13) may be of more practical value, although it must be realized that the quantities of inorganic N arising from the mineralization of organic N in organic manures are more predictable and more consistent under green-house- than under field conditions. In addition, it has become clear from the results obtained on farmers' fields that the quantities in which liquid manure are often applied to soils in the Netherlands, far exceed the quantities needed to ensure a steady, but limited supply of NO_3 to spinach plants.

In general, information obtained on N-nutrition of plants grown in pots, must be viewed with caution when the results are used for predicting the behaviour of plants grown in the field. Since mineralization of soil organic N in pots is necessarily of minor importance, dry-matter production and NO_3-content in leaves are governed mainly by the quantity of N applied and by the age of the plants (Figure 4.9), although environmental factors can also exert an influence (Figure 4.5). As mentioned earlier, under field conditions a number of factors influencing the availability of nitrogen are different. The rooting volume per plant is much larger, and the contribution that minerali-zation of soil organic N makes to the N-nutrition of the plants is also of more importance. On the other hand, leaching of NO_3 from the rooting zone may now be an important factor in lowering N-availability.

Notwithstanding the higher degree of complexity in N-nutrition of spinach plants growing under field conditions, Böhmer [1980] could demonstrate that the yields of spinach and of other vegetables were largely determined by the amounts of plant-available N, being the summation of N-dressing and quantity of inorganic N present in the soil before sowing or planting. She found that maximum yields of spring-grown spinach were attained when 250 kg plant-available N was present in the top 60 cm of the soil profile and when of these 250 kg at least 100 kg was present in the top 30 cm.

With these values reported by Böhmer and the values on available soil N present before sowing, as mentioned in section 5.1.4, the N-dressings with which maximum yields should be obtained in the field experiments of the present study, could be calculated. They were found to agree reasonably closely with the N-dressings needed to obtain maximum yields under field conditions (Table 12, parts a and b).

In Figure 6.1, the quantities of plant-available N (N-dressings plus the amount of available N in the soil before sowing) are plotted against the NO_3-contents in field-grown spinach. As in general the variations in NO_3-contents among replicates within a treatment were large, single values are presented in this figure. The results of experiments conducted in Duiven are presented in A, those collected in Lelystad and Helden in B, and those obtained on farmers' fields in the spring of 1980 (section 5.4) in C.

From Figure 6.1 it can be seen that a relationship exists between the quantity of plant-available N and the NO_3-content in spinach, but that the relationship is not very close, especially not for the farmers' fields (C). An explanation for the aberrant behaviour of the latter is that on the one hand these farmers' fields often receive heavy applications of manure, causing the mineralization of NO_3 during the growth period to be generally larger than observed on the experimental fields (20-50 kg NO_3-N, except for experiment 5), and on the other hand that the sandy texture of the soils of some of these farmers' fields adds a risk of considerable leaching of NO_3 from these soils. Moreover, on the experimental fields, the spinach plants were cut with a knife just below the stem, whereas the samples taken from the farmers' fields were cut 2-3 cm above the soil surface, so that many petioles and older leaves were not included in the harvested material. Since petioles and older leaves are known to be relatively rich in NO_3 (Figure 4.7), the produce harvested from farmers' fields had a lower NO_3-content than that harvested from experimental fields with comparable quantities of plant-available N.

The NO_3-contents presented in Figure 6.1 are expressed on a dry-weight basis. Limits for acceptable NO_3-contents in vegetables are, however, usually expressed on a fresh-weight (FW) basis, in mg NO_3 per kg produce. The data presented in Figure 6.1 can be converted as follows: NO_3-content (in mmol per kg DM) x dry-matter content (in %) x 0.62 = NO_3-content (in mg NO_3 per kg FW). When e.g. the NO_3- and dry-matter contents are 400 mmol per kg DM and 5% respectively, the NO_3-content in mg NO_3 per kg FW is 400 x 5 x 0.62 = 1240.

Recently, Dutch governmental authorities have proposed an upper acceptable limit for NO_3-contents in spinach and endive to be set at 4000 mg per kg FW [Anonymous, 1981]. The aim is, that in the future this limit is lowered to 2500 mg NO_3 per kg FW. In the baby-food industry, the upper acceptable limit for spinach delivered at the factory is momentarily set at 1500 mg NO_3 per kg FW (Nutricia B.V., private communication).

Of all samples presented in Figure 6.1, only one had a NO_3-content exceeding 4000 mg per kg FW. Of the 228 samples taken from experimental fields (Figure 6.1, A and B), 137 (60%) had NO_3-contents lower than 1500, and 206 (90%) lower than 2500 mg NO_3 per kg FW. When the 48 samples of the autumn experiments (4 and 7) are left out of consideration, these percentages become 66% and 97%, respectively.

Of the 120 samples taken from farmers' fields, 81 (68%) had NO_3-contents lower than 1500 and 102 (85%) lower than 2500 mg per kg FW. On a fresh-weight basis, the NO_3-contents in the samples of the experimental field trials conducted in the spring and those in samples of the farmers' fields were of the same magnitude, with comparable quantities of plant-available N.

[82]

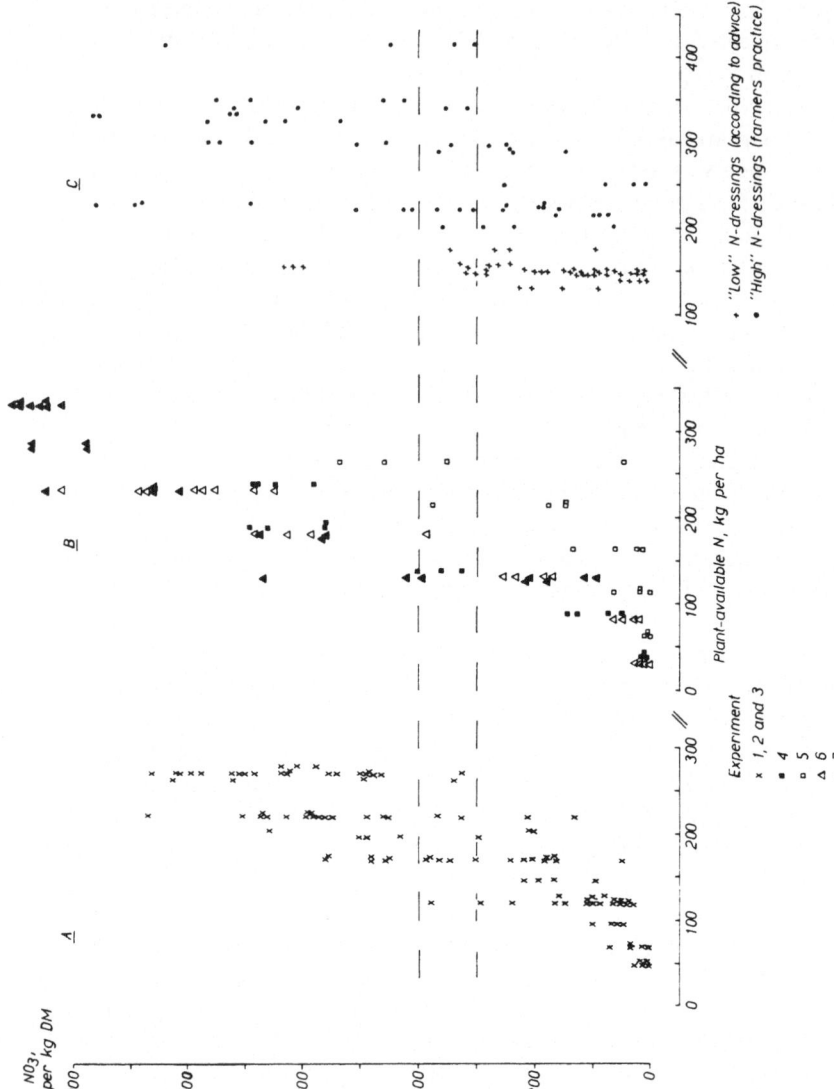

Figure 6.1 Relationship between plant-available N (= N-dressing plus the amount of available N in the soil (0-60 cm) before sowing) and the NO₃-contents of samples from the final harvests of experiments in Duiven (A), Lelystad and Helden (B) and from experiments on farmers' fields (C).

Consequently, all samples of spring experiments were arranged in categories based on quantities of plant-available N (N-dressings plus the amount of available N in the soil before sowing). In Figure 6.2, per category the percentages of samples having NO_3-contents lower than 1500 (A) or 2500 (B) mg per kg FW are shown. From this figure it can be seen that in cases of plant-available N being lower than 150 kg per ha, the NO_3-contents were lower than 1500 mg per kg FW. With 150-175 kg plant-available N, only 12% of the samples had a NO_3-content exceeding 1500 mg per kg FW, whereas the 2500 mg per kg FW limit was not surpassed at all. With higher quantities of plant-available N, the percentage of samples having NO_3-contents exceeding 1500 mg per kg FW increased. With respect to the 2500 mg per kg FW limit, it can be seen that this limit was exceeded considerably when quantities of plant-available N rose above 225 kg per ha.

From these results, it can be concluded that with quantities of plant-available N being less than 175 and 225 kg per ha, spinach grown in spring will have NO_3-contents below the limits of 1500 and 2500 mg per kg FW, respectively. With the former quantity of plant-available N, yields will be 10-20% lower than the maximally attainable ones, whereas with the latter quantity the yield depression will be about 5%. It must be emphasized that these predictions are based on findings obtained with relatively small numbers of samples. Further research on this subject was carried out in 1982 by the Institute of Soil Fertility at Haren. Results obtained in field experiments conducted in the spring of 1980, and preliminary results of field experiments conducted in the spring of 1981 agreed fairly well with those obtained in the present study, especially for spring-grown spinach on sandy soils (J.H.Pieters, unpublished). For spinach grown on clay soils, the limits for NO_3-contents mentioned earlier, were surpassed at quantities of plant-available N exceeding those found in the present study [van der Boon & Pieters, 1981].

So far, three measures aimed at reducing the NO_3-content in spinach have been discussed. These were, (1) inducing the spinach plants to absorb a portion of the N as NH_4 through the use of a combination of NH_4- and NO_3-fertilizers and of nitrification inhibitors, (2) tempering the availability of N through the use of organic manures as N-source, and (3) reducing the quantities of NO_3-N to be applied. A fourth measure to be taken concerns variations to be imposed in the timing of N-applications.

In soil cultures of the present investigation, variations in timing of N-application were found not to affect the NO_3-contents and yields of spinach (Table 6). In the field, both NO_3-contents and yield levels at intermediate harvests were suppressed when the N-fertilizer was applied in the form of split dressings, but at the final harvest little of these effects had remained (Table 13).

In a number of experiments conducted elsewhere, splitting the application of N resulted in higher NO_3-contents in spinach [Kuhlen, 1962; Maga et al., 1976; Mehwald, 1973; Nicolaisen & Zimmermann, 1948; Zimmermann,

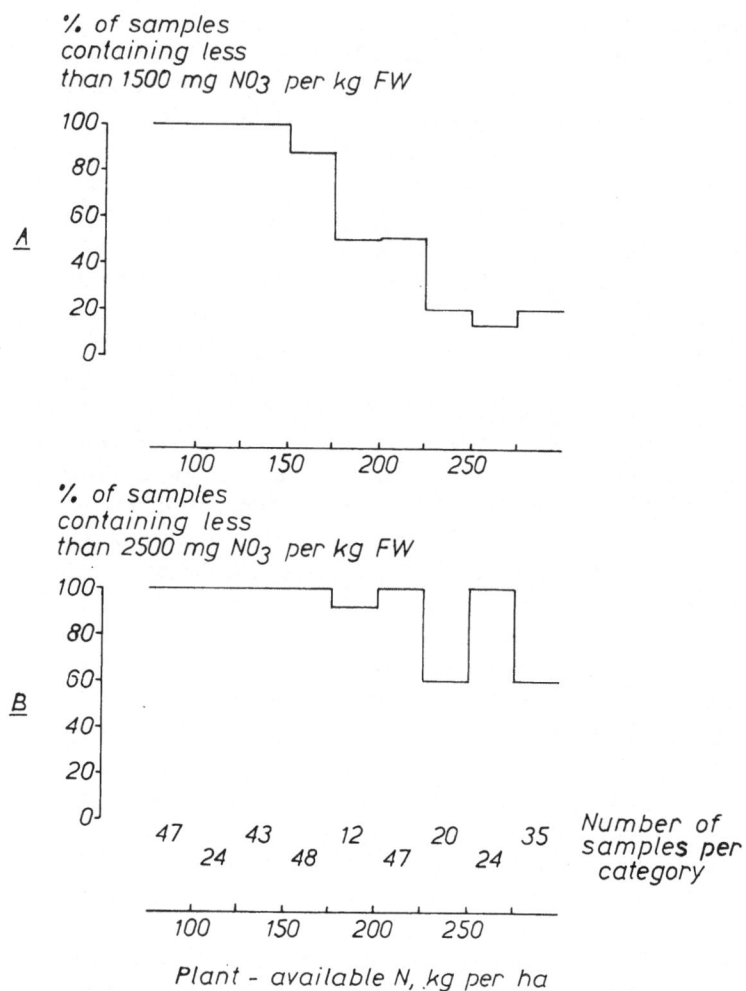

% of samples
containing less
than 1500 mg NO₃ per kg FW

% of samples
containing less
than 2500 mg NO₃ per kg FW

Plant - available N, kg per ha

Figure 6.2 Percentage of spinach samples from final harvests of spring field experiments and from experiments on farmers' fields containing (A) less than 1500 mg and (B) less than 2500 mg NO₃ per kg fresh weight for various categories of plant-available N (= N-dressing plus the amount of available N in the soil (0-60 cm) before sowing).

1966]. Barker et al. [1971], however, observed that a split application of N led to a decrease in NO₃-content in spinach. In the same study, it was observed that side dressings of ammonium nitrate or urea resulted in lower NO₃-contents than did those of potassium nitrate. Similar results were obtained in the present study (Table 14).

In general, however, it must be remarked that in view of existing variabilities in field conditions, such as soil texture and soil-N status, and in Dutch weather conditions with an ensuing uncertainty about harvest dates, it

[84]

appears practically impossible to arrive at a justified decision on the dates at which second and third N-dressings should be applied. Consequently, the conclusion must be that for a crop like spinach, having a short growing season (40-60 days) and forming the major portion of its dry matter in the last few weeks before harvest, any partitioning of the quantity of nitrogen to be applied does not seem to be realistic and, hence, not advisable.

When all results presented in the aforegoing are reviewed, the conclusion can be drawn that in the Netherlands certainly opportunities are available to grow spinach with an acceptably low NO_3-content in a profitable way, provided some simple and cheap measures are taken that have appeared to be effective in preventing the accumulation of NO_3 in the leaves. To help growers in reaching the goal of producing healthy spinach, the following recommendations can be made:

1. The quantity of NO_3 present in the top 60 cm of the soil at the start of the growing season should not be larger than 225 kg N per ha. With this quantity of available N, the decline in yield, relative to the maximally attainable yield, usually does not exceed the level of 5%.
2. The spinach should be preferably grown in the spring season.
3. The pH of the soil should not be lower than pH-H_2O 6.
4. The spinach should not be harvested prematurely.
5. The harvest should preferably take place on a cool, bright day.
6. Harvesting should be carried out in such a way that a portion of the petioles and older leaves remain in the field.
7. When, in the future, unequivocal and consistent differences in NO_3-contents of leaves among spinach varieties might be shown to exist, spinach growers should give preference to such varieties.
8. When in advance it is known that certain of these above recommendations cannot be met, the quantity of available NO_3-N in the soil should be lowered to a level of 175 kg N per ha. The yield decline arising from this reduction in available N will usually not be larger than 25%.

Summary

Ingestion of high amounts of nitrate by man can be considered hazardous to human health. In the human body, nitrate can be reduced to nitrite which may cause methemoglobinemia. Furthermore, the possible formation of N-nitroso-compounds from nitrite and secondary nitrogen compounds in the human stomach constitutes a risk, as for laboratory animals many N-nitroso-compounds have been shown to act as potent carcinogens.

In the Netherlands, acceptable daily intakes of nitrate and nitrite, as set by the FAO/WHO for additives, are sometimes exceeded. The intakes of nitrate and nitrite therefore should be reduced. As nitrite intake is linked to nitrate intake which for about 75% is of vegetable origin, the nitrate contents of vegetables should be suppressed as much as possible.

Spinach is one of the vegetables frequently having a high nitrate content. In the Netherlands, most spinach is grown for the processing industry and a portion of the processed spinach is used as baby food.

Many environmental factors and cultural measures can affect the nitrate content in spinach. A review of relevant literature is presented in chapter 2. In the present study, some of these environmental factors have been invest-igated with respect to their influences on nitrate content and yield of spin-ach. Most attention, however, has been given to cultural measures, especial-ly those concerning application of nitrogen fertilizers.

The materials and methods used in this study are discussed in chapter 3.

In chapter 4, the results of experiments carried out indoors, with water-, sand- and soil-cultures, are described. In water-culture experiments, the cation:anion uptake ratio for spinach supplied with nitrate as sole nitrogen source, proved to be close to unity. When nitrogen (NO_3) was withheld from the spinach plants over a 3- or 9-day period, the nitrate contents in the aerial parts significantly decreased (section 4.2).

In sand-culture experiments spinach showed very poor growth when sup-plied with ammonium-nitrogen only. With both ammonium- and nitrate-nitrogen added, the nitrate contents in spinach proved to be not always lower than with nitrate-nitrogen only. On a whole-plant basis, the carboxylate (C-A) : organic nitrogen ratio was found to reflect the extent of nitrate- or ammonium nutrition fairly closely (section 4.3).

In soil-culture experiments both the nitrate contents and dry-matter yields of spinach were found to be strongly affected by the amount of nitrate ap-plied. Yields usually showed a maximum, whereas nitrate contents did not. The nitrate contents in petioles of spinach were considerably higher than in laminae, and nitrate accumulated mainly in older leaves. In most soil-culture experiments, the nitrate contents in the aerial parts of spinach decreased with increasing age of the plants. Nitrate contents in autumn-grown spinach were higher than those in spinach grown in spring. Variations in timing of application of nitrate-nitrogen did not affect the nitrate contents. Compared with a normal soil-moisture content, a low soil-moisture content was found to increase, and a high soil-moisture content was found not to affect the nitrate content in spinach. Liming a sandy soil, resulting in increased soil pH-values, caused the nitrate contents in spinach to decline. Molybdenum applied as a spray onto the leaves of spinach, grown on a sandy as well as on a clay-loam soil, did not have any effect on the nitrate contents in the aerial parts, but differences in soil type appeared to affect these contents (section 4.4).

In a soil-culture experiment with a sandy soil, NH_4-N dressings produced higher nitrate contents and lower dry-matter yields of spinach than did NO_3-N-dressings. With the use of a nitrification inhibitor (DCD), however, lower nitrate contents were found with NH_4-N-dressings. With a clay-loam soil used in a growth-chamber experiment, replacement of about 30% of the NO_3-N by NH_4-N, with a nitrification inhibitor (nitrapyrin) added, did not result in a decrease in nitrate content, but replacement of about 60% of the NO_3-N by NH_4-N did. In the same experiment, a decrease in light intensity from 70 to 33 W per m^2 and a rise in temperature from 12 to 22°C caused the nitrate content in spinach to increase (section 4.5).

In a comparison of different N-carriers in a soil culture, positive results were obtained with sulphur-coated urea (section 4.6), with farmyard manure and pig-manure slurry (section 4.7), when plant-available N was taken into account. Variations in P-dressings as well as in soil P-status did not affect the nitrate content and yield of spinach (section 4.8). K-dressings in general increased the nitrate contents and yields of spinach, with K_2SO_4 more than KCl being responsible for increases in nitrate contents (section 4.9).

Large differences among spinach varieties were found with respect to nitrate contents in leaves (section 4.10).

In chapter 5, the results of experiments conducted outdoors are described. In the field experiments, yield, NO_3- and N(total)-contents and total N in the aerial parts of spinach increased with increasing amounts of NO_3-N applied. With the highest NO_3-dressings, NO_3-N accounted for 9-27% of the total N in spinach. The corresponding NO_3-contents ranged from 300 to 1100 mmol per kg DM. In one experiment, varietal differences in NO_3-content could be attributed to differences in age of the crop. Differences in NO_3-contents between 'morning' and 'evening' harvests were found only on a bright day with high irradiation and with high NO_3-levels in the plants.

Available N in the soil profile before sowing ranged from 32 to 72 kg N per ha in the five spring experiments and from 40 to 132 kg N per ha in the two autumn experiments. In general, the net mineralization of organic N during the experimental periods was about 35 kg N per ha. Nitrogen appeared to be taken up by spinach mainly from the top 60 cm of the soil profile.

[87]

Of the plant nutrients other than N only the contents of SO_4 and H_2PO_4 were not affected by variations in NO_3-dressing (section 5.1).

Variations in timing of nitrogen applications affected yields in one, but not in another field experiment. NO_3- and N(total)-contents and total N in the aerial parts of spinach in general decreased due to a partitioning of total N applied in one field experiment, whereas in another one, top-dressed N applied as NH_4 or $NH_4 + NO_3$ resulted in lower values than did comparable dressings of NO_3 (section 5.2).

Due to partial or complete replacement of NO_3-N by NH_4-N, yields, NO_3- and N(total)-contents, total N and the ratio (C-A):organic N in the aerial parts of spinach decreased in three field experiments, the effects in general being more pronounced with than without a nitrification inhibitor added. In two other field experiments, the effects of variations in NO_3:NH_4-ratio were much less pronounced or absent (section 5.3).

On farmers' fields with 'low' N-dressings, plant-available N (= N-dressings plus the amount of available N in the soil before sowing) ranged from 135 to 175 kg N per ha, while with 'high' N-dressings plant-available N ranged from 200-415 kg N per ha. Compared with 'high' N-dressings, yields with 'low' N-dressings on the average were 17% lower, while NO_3-contents were 64% lower (section 5.4).

A discussion of the results is presented in chapter 6. Measures aimed at utilizing or manipulating environmental and soil conditions, at exploiting varietal differences and those aimed at regulating nutritional conditions, other than the ones governed by nitrogen, with the purpose of growing spinach with an economically acceptable yield level and with a low NO_3-content, are dealt with in section 6.1.

Experimentation aimed at finding the proper amount, the proper form and the proper timing of application of N for attaining the above-mentioned goal, is discussed in section 6.2. In the latter, special attention is paid to the relationship between the quantity of plant-available N (= N-dressing plus the amount of available N in the soil before sowing) and the NO_3-content in spinach leaves. Based on data from spring field experiments and from experiments on farmers' fields, the amounts of plant-available N with which critical levels of NO_3-contents are not surpassed, are presented. Recommendations for practical use in commercial spinach growing are formulated in the final portion of section 6.2.

List of figures

List of tables

[91]

Appendix

NO_3-contents in mg NO_3 per kg FW calculated from NO_3-contents expressed as m mol per kg DM with various dry-matter contents.

NO_3 (m mol per kg DM)	100	200	300	400	500	600	700	800	900	1000
Dry matter content (%)										
	mg NO_3 per kg FW:									
4	248	496	744	992	1240	1488	1736	1984	2232	2480
6	372	744	1116	1488	1860	2232	2604	2976	3348	3720
8	496	992	1488	1984	2480	2976	3472	3968	4464	4960
10	620	1240	1860	2480	3100	3720	4340	4960	5580	6200

Conversion scheme for spinach with a dry-matter content of 8%.

% NO_3-N in DM	mmol NO_3 per kg DM	mg NO_3-N per kg FW	mg NO_3 per kg FW	mg KNO_3 per kg FW
1.40	= 1000	= 1120	= 4960	= 8080
1.25	= 893	= 1000	= 4429	= 7214
0.28	= 202	= 226	= 1000	= 1629
0.17	= 124	= 139	= 614	= 1000

References

1. Acar I and Ahrens E (1978) Zum Problem des Nitrat- und Nitritgehaltes bei Gemüse, insbesondere Spinat. 1. Einfluss der Stickstoffdüngung und der Verarbeitung zu Gefrierspinat. Chem Mikrobiol Technol Lebensm 5, 170-174
2. Achtzehn MK and Hawat H (1969) Die Anreicherung von Nitrat in den Gemüsenarten - eine Möglichkeit der Nitratintoxikation bei Säuglingen. Nahrung 13, 667-676 .
3. Achtzehn MK and Hawat H (1971) Einfluss industrieller Vorbehandlung auf den Gehalt essentieller und nicht essentieller Inhaltsstoffe im Spinat. Nahrung 15, 527-537
4. Aldershoff WG (1982) Nitraat in gewassen en onze gezondheid. Bedrijfsontwikkeling 13, 273-280
5. Allison FE (1973) Soil organic matter and its role in crop production. New York: Elsevier Scientific Publishing Co
6. Anonymous (1981) Naar minder nitraat in bladgroenten. Persbericht Ministerie van Landbouw en Visserij No. 442, 7 december 1981
7. Armijo R and Coulson AH (1975) Epidemiology of stomach cancer in Chile - the role of nitrogen fertilizers. Int J Epidemiol 4, 301-309
8. Aslam M, Oaks A and Huffaker RC (1976) Effect of light and glucose on the induction of nitrate reductase and on the distribution of nitrate in etiolated barley leaves. Plant Physiol (Bethesda) 58, 588-591
9. Aworh OC, Hicks JR, Minotti PL and Lee C (1980) Effects of plant age and nitrogen fertilization on nitrate accumulation and postharvest nitrite accumulation in fresh spinach. J Am Soc Hortic Sci 105, 18-20
10. Barker AV (1975) Organic vs. inorganic nutrition and horticultural crop quality. HortScience 10, 50-53.
11. Barker AV and Maynard DN (1971) Nutritional factors affecting nitrate accumulation in spinach. Commun Soil Sci Plant Anal 2, 471-478
12. Barker AV and Mills HA (1980) Ammonium and nitrate nutrition of horticultural crops. Hortic Rev 2, 395-423
13. Barker AV, Maynard DN and Mills HA (1974) Variations in nitrate accumulation among spinach cultivars. J Am Soc Hortic Sci 99, 132-134
14. Barker AV, Peck NH and MacDonald GE (1971) Nitrate accumulation in vegetables. I. Spinach grown in upland soils. Agron J 63, 126-129
15. Becker KF (1965) Nitrat- and Nitritgehalt in Spinat. Bundesgesundheitsblatt 8, 246-248
16. Beevers L and Hageman RH (1969) Nitrate reduction in higher plants. Annu Rev Plant Physiol 20, 495-522
17. Beevers L and Hageman RH (1972) The role of light in nitrate metabolism in higher plants. Photophysiology 7, 85-113
18. Bengtsson BL (1968) Effect of nitrification inhibitor on yield and nitrate content of spinach. Z Pflanzenernaehr Bodenkd 121, 1-4
19. Ben Zioni A, Vaadia Y and Lips SH (1971) Nitrate uptake by roots as regulated by nitrate reduction products of the shoot. Physiol Plant 24, 288-290
20. Bodiphala T and Ormrod DP (1971) Factors affecting the nitrate content of vegetable and fruit foods. Can Inst Food Technol J 4, 6-8
21. Boek K and Schuphan W (1959) Der Nitratgehalt von Gemüsen in Abhängigkeit von Pflanzenart und einigen Umweltfaktoren. Qual Plant Mat Veg 5, 199-208

22. Böhmer M (1980) Der Mineralstickstoffgehalt von Böden mit Feldgemüsebau und seine Bedeutung für die Stickstoffernährung der Pflanze. Dissertation Universität Hannover

23. Boon J van der and Pieters JH (1981) Stikstofaanbod uit grond en bemesting, en nitraat in spinazie. Stikstofbemestingsproeven bij spinazie op zeeklei in 1980. Nota 97, Instituut voor Bodemvruchtbaarheid te Haren

24. Breimer T and Slangen JHG (1981) Pretreatment of soil samples before NO3-N analysis. Neth J Agric Sci 29, 15-22

25. Breteler H (1973) A comparison between ammonium and nitrate nutrition of young sugar-beet plants grown in nutrient solutions at constant acidity. 1. Production of dry matter, ionic balance and chemical composition. Neth J Agric Sci 21, 227-244

26. Briggs GG (1975) The behaviour of the nitrification inhibitor "N-Serve" in broadcast and incorporated applications to soil. J Sc Food Agric 26, 1083-1092

27. Brink NDH van de, Coster DRA and Drost-de Wijs H (1968) Nitraat en nitriet in spinazie. Voeding en Techniek 2, 1122-1126

28. Brown JR and Smith GE (1967) Nitrate accumulation in vegetable crops as influenced by soil fertility practices. Missouri Agric Exp Stn Res Bull 920

29. Brown JR, Lambeth VN and Blevins DG (1969) Nutrient interaction effects on yield and chemical composition of spinach and green beans. Missouri Agric Exp Stn Res Bull 963

30. Buishand Tj (1974) Teelt van spinazie. Consulentschap in Algemene Dienst voor de Groenteteelt in de Vollegrond in Nederland te Alkmaar

31. Burg PFJ van, Groot EH, Keller GHM and Rauw GJG (1967) De stikstofbemesting van spinazie in verband met opbrengst en kwaliteit. Onderzoek 1967. Versl Landbouwkd Bur Ned Stikstofmestst Ind no B 163

32. Burg PFJ van, Groot EH, Keller GHM and Rauw GJG (1969) Stikstofbemesting spinazie, 1968. Versl Landbouwkd Bur Ned Stikstofmestst Ind no B 187

33. Cantliffe DJ (1972-1) Nitrate accumulation in spinach grown under different light intensities. J Am Soc Hortic Sci 97, 152-154

34. Cantliffe DJ (1972-2) Nitrate accumulation in vegetable crops as affected by photoperiod and light duration. J Am Soc Hortic Sci 97, 414-418

35. Cantliffe DJ (1972-3) Nitrate accumulation in spinach grown at different temperatures. J Am Soc Hortic Sci 97, 674-676

36. Cantliffe DJ (1973-1) Nitrate accumulation in table beets and spinach as affected by nitrogen, phosphorus, and potassium nutrition and light intensity. Agron J 65, 563-565

37. Cantliffe DJ (1973-2) Nitrate accumulation in spinach cultivars and plant introductions. Can J Plant Sci 53, 365-367

38. Cantliffe DJ and Phatak SC (1974-1) Effect of herbicides on weed control and nitrate accumulation in spinach. HortScience 9, 470-472

39. Cantliffe DJ and Phatak SC (1974-2) Nitrate accumulation in greenhouse vegetable crops. Can J Plant Sci 54, 783-788

40. Cantliffe DJ, MacDonald GE and Peck NH (1974) Reduction of nitrate accumulation by molybdenum in spinach grown at low pH. Commun Soil Sci Plant Anal 5, 273-282

41. Coïc Y, Lesaint Ch and LeRoux F (1961) Comparison de l'influence de la nutrition nitrique et ammoniacale combinée ou non avec une déficience en acide phosphorique, sur l'absorption et le métabolisme des anions-cations et plus particulièrement des acides organiques chez le maïs. Comparison du maïs et de la tomate quant à l'effet de la nature de l'alimentation azotée. Ann Physiol Vég (Paris) 3, 141-163

42. Coïc Y, Lesaint Ch and LeRoux F (1962) Effets de la nature ammoniacale ou nitrique de l'alimentation azotée et du changement de la nature de cette alimentation sur le métabolisme des anions et cations chez la tomate. Ann Physiol Vég (Paris) 4, 117-125

43. Cools MH, Meijs MQ van der, Roorda van Eysinga JPNL and Stolk JH (1980) Het nitraatgehalte van enkele spinazierassen geteeld onder glas. Intern verslag no. 1. Proefstation voor Tuinbouw onder Glas te Naaldwijk, januari 1980

44. Corré WJ and Breimer T (1979) Nitrate and nitrite in vegetables. Wageningen: Centre for Agricultural Publishing and Documentation

45. Cuello C, Correa P, Haenszel W, Gordillo G, Brown C, Archer M and Tannenbaum S (1976) Gastric cancer in Colombia. I. Cancer risk and suspect environmental agents. J Natl Cancer Inst 57, 1015-1020

46. Descamps P (1972) Les nitrates dans les légumes de conserve: cas des épinards et des carottes. Centre Technique des Conserves de Produits Agricoles (CTCPA). Station Experimentale de Dury

47. Dijkshoorn W (1969) The relation of growth to the chief ionic constituents of the plant. In: Rorison IH, ed. Ecological aspects of the mineral nutrition of plants, pp 201-213. Oxford: Blackwell Scientific Publications

48. Dijkshoorn W and Wijk AL van (1967) The sulphur requirements of plants as evidenced by the sulphur-nitrogen ratio in the organic matter - a review of published data. Plant Soil 26, 129-157

49. Dressel J (1976) Abhängigkeit qualitätsbeeinflussender pflanzlicher N-haltiger Inhaltstoffe von der Düngungsintensität. Landwirtsch Forsch 33 II SH, 326-334

50. Dressel J and Jung J (1970) Der Einfluss der Düngung auf verschiedene Inhaltstoffe von Spinat. Ernaehr Umsch 17, 524-527

51. Eerola M, Varo P and Koivistoinen P (1974) Nitrate and nitrite in spinach (Spinacia oleracea L.) as affected by application of different levels of nitrogen fertilizer, blanching, and storage after thawing of frozen product. Acta Agric Scand 24, 286-290

52. Egmond F van (1971) Inorganic cations and carboxylates in young sugarbeet plants. Proc 8th Colloq Int Potash Inst (Uppsala), pp 104-117

53. Egmond, F van (1975) The ionic balance of the sugar-beet plant. Agric Res Rep 832. Wageningen: Centre for Agricultural Publishing and Documentation

54. Egmond, F van (1978) Critique of "Absorption and utilization of ammonium nitrogen by plant". Nitrogen nutritional aspects of the ionic balance of plants. In: Nielsen DR and MacDonald JG, eds. Nitrogen in the environment 2, pp 171-189. New York: Academic Press

55. Fraser P, Chilvers C, Beral V and Hill MJ (1980) Nitrate and human cancer: a review of the evidence. Int J Epidemiol 9, 3-11

56. Frota JNE and Tucker TC (1972) Temperature influence on ammonium and nitrate absorption by lettuce. Soil Sci Soc Am Proc 36, 97-100

57. Gasser JKR (1970) Nitrification inhibitors - their occurrence, production and effects of their use on crop yields and composition. Soils Fert 33, 547-554

58. Geleperin MD, Moses VJ and Fox G (1976) Nitrate in water supplies and cancer. Ill Med J 149, 251-253

59. Geyer B (1978) Untersuchungen zur Wirkung hoher Stickstoffgaben auf den Nitratgehalt von Freilandgemüse. Arch Gartenbau 26, 1-13

60. Grujić S and Kastori R (1974) Einfluss der verschiedenen Mineralstoffernährung auf den Nitrat- und Nitritgehalt im Spinat. Proc IV Int Congress Food Sci Technol (Madrid) 3, 272-277

61. Haenszel W, Kurihara M, Segi M and Lee RKC (1972) Stomach cancer among Japanese in Hawaii. J Natl Cancer Inst 49, 969-988

62. Haenszel W, Kurihara M, Locke FB, Shinuzu K and Segi M (1976) Stomach cancer in Japan. J Natl Cancer Inst 56, 265-274
63. Hansen H (1978) The influence of nitrogen fertilization on the chemical composition of vegetables. Qual Plant 28, 45-63
64. Harada M, Ishiwata H, Nakamura Y, Tanimura A and Ishidate M (1975) Changes of nitrite and nitrate concentrations in human saliva after ingestion of salted Chinese cabbage. J Food Hyg Soc Jpn 16, 11-18
65. Haynes RJ and Goh KM (1978) Ammonium and nitrate nutrition of plants. Biol Rev Camb Philo Soc 53, 465-510
66. Hewitt EJ (1975) Assimilatory nitrate-nitrite reduction. Annu Rev Plant Physiol 26, 73-100
67. Hiatt AJ (1967) Relationship of cell sap pH to organic acid change during ion intake. Plant Physiol (Bethesda) 42, 294-298
68. Hildebrandt (1976) Zur Problematik der Nitrosamine in der Pflanzenernährung. Dissertation Justus Liebig Universität Giessen
69. Hill MJ, Hawksworth GM and Tattersall G (1973) Bacteria nitrosamines and cancer of the stomach. Br J Cancer 28, 562-567
70. Houba VJG, Egmond F van and Wittich EM (1971) Changes in production of organic nitrogen and carboxylates (C-A) in young sugar-beet plants grown in nutrient solutions of different nitrogen compositions. Neth J Agric Sci 19, 39-47
71. Huber DM and Watson RD (1974) Nitrogen form and plant disease. Annu Rev Phytopathol 12, 139-165
72. Huffaker RC and Rains DW (1978) Factors influencing nitrate acquisition by plants; assimilation and fate of reduced nitrogen. In: Nielsen DR and MacDonald JG, eds. Nitrogen in the environment 2, pp 1-43. New York: Academic Press
73. Hulewicz D and Mokrzecka E (1971) Ertragsabhängigkeit des Spinats und einiger seiner Wertbestimmenden Bestandteile von der Düngung. Z Pflanzenernaehr Bodenkd 130, 214-224
74. Ishiwata H, Boriboon P, Nakamura Y, Harada M, Tanimura A and Ishidate M (1975) Studies on in vivo formation of nitroso compounds. II. Changes of nitrite and nitrate concentration in human saliva after ingestion of vegetables or sodium nitrate. J Food Hyg Soc Jpn 16, 19-24
75. Jackson WA (1978) Critique- of "Factors influencing nitrate acquisition by plants: assimilation and fate of reduced nitrogen". Nitrate acquisition and assimilation by higher plants: processes in the root system. In: Nielsen DR and MacDonald JG, eds. Nitrogen in the environment 2, pp 45-88. New York: Academic Press
76. Jacobson L and Ordin L (1954) Organic acid metabolism and ion absorption in roots. Plant Physiol (Bethesda) 29, 70-75
77. Jacquin F and Papadopoulos G (1977) Influence de la forme de fumure azotée sur l'accumulation des nitrates dans des plants d'épinards cultivés en vases de végétation. Bull Ec Nat Super Agron Ind Aliment 19, 101-104
78. Joint FAO/WHO Expert Committee on Food Additives (JECFA)(1974) Toxicological evaluation of certain food additives with a review of general principles and specifications. 17th Report JECFA, FAO Nutrition Meetings Report Series No 53; WHO Technical Report Series No 539
79. Joint FAO/WHO Expert Committee on Food Additives (JECFA)(1976) Evaluation of certain food additives. 20th Report JECFA, WHO Technical Report Series No 599; FAO Food and Nutrition Series No 1
80. Joint Iran-International Agency for Research on Cancer Study Group (1977) Oesophageal cancer studies in the Caspian Littoral of Iran: Results of population studies - a prodrome. J Natl Cancer Inst 59, 1127-1138
81. Jung J and Dressel J (1978) Umsetzungsvorgänge und Inhibierungsmöglichkeiten

bei Boden- und Düngerstickstoff. Landwirtsch Forsch 34 II SH, 74-89

82. Jurkowska H (1971) Effect of dicyanodiamide on the content of nitrates and oxalic acid in spinach. Agrochimica 15, 445-453

83. Jurkowska H and Wojciechowicz T (1974) Investigations on the content of oxalic acids and nitrates in plants. Part II. Effect of dicyanodiamide as an inhibitor of nitrification on the accumulation of oxalic acid and nitrates in plants as depending on the nitrogen doses (in Polish). Acta Agrar Silvestria Ser Agrar 14/2, 15-24

84. Kerkhoff P and Slangen JHG (1980) Nitrificatie-remstoffen in land- en tuinbouw. Interne Mededeling no 54, Vakgroep Bodemkunde en Bemestingsleer, Landbouwhogeschool, Wageningen

85. Kick H and Massen GG (1973) Der Einfluss von Dicyandiamid und N-Serve in Verbindung mit Ammoniumsulfat als N-dünger auf die Nitrat-und Oxalsäuregehalte von Spinat (Spinacia oleracea). Z Pflanzenernaehr Bodenkd 135, 220-226

86. Kirkby EA (1969) Ion uptake and ionic balance in plants in relation to the form of nitrogen nutrition. In: Rorison IH, ed. Ecological aspects of the mineral nutrition of plants, pp 215-255. Oxford: Blackwell Scientific Publications

87. Kirkby EA (1974) Recycling of potassium in plants considered in relation to uptake and organic acid accumulation. In: Wehrmann J, ed. Proc 7th Int Colloq Plant Anal Fert Probl (Hannover) 2, 557-568

88. Kirkby EA and Knight AH (1977) Influence of the level of nitrate nutrition on ion uptake and assimilation, organic acid accumulation, and cation-anion balance in whole tomato plants. Plant Physiol (Bethesda) 60, 349-353

89. Klett M (1968) Untersuchungen über Licht- und Schattenqualität in Relation zum Anbau und Test von Kieselpreparaten zur Qualitätshebung. Institut für biologisch-dynamische Forschung, Darmstadt

90. Knauer N (1970) Beeinflussung der Qualität von Spinat durch pflanzenbauliche Massnahmen. Ernaehr Umsch 17, 5-8

91. Knauer N and Simon C (1968) Uber den Einfluss der Stickstoffdüngung auf den Ertrag sowie Nitrat-, Mineralstoff- und Oxalsäuregehalt von Spinat. Z Acker Pflanzenb 128, 197-220

92. Kübler W and Simon C (1969) Nitrat-Nitritvergiftigungen im Säuglingsalter. Nutr Dieta 11, 111-119

93. Kuhlen H (1962) Der Nitratgehalt von Spinat im Abhängigkeit von der Stickstoffdüngung und andere ökologische Faktoren. Proc XVIth Hortic Congress 2, 216-222

94. Lambeth VN, Fields ML, Brown JR, Regan WS and Blevins DG (1969) Detinning by canned spinach as related to oxalic acid, nitrates and mineral composition. Food Technol 23, 840-842

95. Lee CY, Shallenberger RS, Downing DL, Stoewsand ES and Peck NH (1971) Nitrate and nitrite nitrogen in fresh, stored and processed table beets and spinach from different levels of field nitrogen fertilization. J Sci Food Agric 22, 90-92

96. Loggers G (1979) Nitraat, overdaad schaadt. Voeding (the Hague) 40, 431-433

97. Lorenz OA and Weir BL (1974) Nitrate accumulation in vegetables. In: White PL and Robbins D, eds. Environmental quality and food supply, pp 93-103. New York: Futura Publications

98. Lycklama JC (1963) The absorption of ammonium and nitrate by perennial rye-grass. Acta Bot Neerl 12, 361-423

99. Maercke D van (1973) Stikstofbemesting en het nitraatgehalte van spinazie. Meded Fac Landb Wetensch Rijks Univ Gent 38, 486-503

100. Maercke D van and Vereecke M (1976) Morfo-fysiologische en chemische studie van enkele spinazierassen. Landbouw Tijdschr 29, 1235-1254

101. Maga JA, Moore FD and Oshima N (1976) Yield, nitrate levels and sensory

properties of spinach as influenced by organic and mineral nitrogen fertilizer levels. J Sci Food Agric 27, 109-114

102. Magee PN and Barnes JM (1967) Carcinogenic nitroso compounds. Adv Cancer Res 10, 163-246

103. Maynard DN and Barker AV (1971) Critical nitrate levels for leaf lettuce, radish, and spinach plants. Commun Soil Sci Plant Anal 2, 461-470

104. Maynard DN and Barker AV (1972) Nitrate content of vegetable crops. HortScience 7, 224-226

105. Maynard DN and Barker AV (1974) Nitrate accumulation in spinach as influenced by leaf type. J Am Soc Hortic Sci 99. 135-138

106. Maynard DN and Barker AV (1979) Regulation of nitrate accumulation in vegetables. Acta Hortic 93, 153-159

107. Maynard DN and Lorenz OA (1979) Controlled release fertilizers for horticultural crops. Hortic Rev 1, 79-140

108. Maynard DN, Barker AV, Minotti PL and Peck NH (1976) Nitrate accumulation in vegetables. Adv Agron 28, 71-118

109. Mehwald J (1973) Die Düngung bei Spinat im Frühjahrs- und Herbstanbau. Gemüse 9, 197-198

110. Mengel K and Kirkby EA (1978) Principles of plant nutrition. Bern: International Potash Institute

111. Merkel D (1975) Oxalatgehalt und Kationen-Anionen-Gleichgewicht von Spinat-pflanzen in Abhängigkeit vom $NO_3 : NH_4$-Verhältnis in der Nährlösung. Land-wirtsch Forsch 28, 34-40

112. Meyers RJK and Paul EA (1968) Nitrate ion electrode method for soil nitrate nitrogen determination. Can J Soil Sci 48, 369-371

113. Mills HA, Barker AV and Maynard DN (1976) Effects of nitrapyrin on nitrate accumulation in spinach. J Am Soc Hortic Sci 101, 202-204

114. Minotti PL and Stankey DL (1973) Diurnal variation in the nitrate concentration of beet. HortScience 8, 33-34

115. Mirvish SS (1977) N-nitroso compounds: Their chemical and in vivo formation and possible importance as environmental carcinogens. J Toxicol Environ Health 2, 1267-1277

116. Mol HJ (1979) Preventie van schadelijke stoffen in tuinbouwprodukten. Bedrijfs-ontwikkeling 10, 948-954

117. Moore FD (1973) N-serve nutrient stabilizer: A nitrogen management tool for leafy vegetables. Down to Earth 28, 4-7

118. Moore FD, Riggert CE and Holbrook TB (1977) Effect of fertilizer nitrogen nitrification suppression on spinach in alkaline soil. HortScience 12, 412

119. Nicolaisen W and Zimmermann H (1968) Der Einfluss der Stickstoffdüngung auf den Nitratgehalt von Spinat unter wechselnden klimatischen Bedingungen. Garten-bauwissenschaft 33, 353-380

120. Novozamsky I, Eck R van, Schouwenburg JCh van and Walinga I (1974) Total nitrogen determination in plant material by means of the indophenolblue method. Neth J Agric Sci 22, 3-5

121. Nurzynski J (1976) Effect of the chloride and sulphate form of potassium on the quantitative and qualitative aspects of the yields of some vegetable crops on garden peat (in Polish). Biul Warzywniczy 19, 105-118

122. Øien A and Selmer-Olsen AR (1969) Nitrate determination in soil extracts with the nitrate electrode. Analyst 94, 888-894

123. Olday FC, Barker AV and Maynard DN (1976) A physiological basis for different patterns of nitrate accumulation in two spinach cultivars. J Am Soc Hort Sci 101, 217-219

124. Pavlek P, Durman P, Heneberg R and Horgas D (1974) Effect of some climatic factors upon the yield and dry matter and nitrate content of the studied spinach varieties (in Polish). Poljopr Znan Smotra 33, 133-141

125. Pflüger R and Wiedemann R (1977) Der Einfluss monovalenter Kationen auf die Nitratreduktion von Spinacia oleracea L. Z Pflanzenphysiol 85, 125-133

126. Prasad R, Rajale GB and Lakhdive BA (1971) Nitrification retarders and slow-release nitrogen fertilizers. Adv Agron 23, 337-383

127. Preussmann R (1981) Nitrat in der Nahrung - ein Gesundheitsrisiko? In: Vorträge der Informationstagung 'Nitrat in Gemüsebau und Landwirtschaft', 23. November 1981, Gottlieb Duttweiler-Institut, Rüschlikon/ Zürich, pp 17-35

128. Prummel J (1971) Fosfaat- en kalibemesting van bladspinazie en stamslabonen op landbouwgronden. Bedrijfsontwikkeling 2 (editie Tuinbouw), 49-54

129. Rao KP and Rains DW (1976) Nitrate absorption by barley. I. Kinetics and energetics. Plant Physiol (Bethesda) 57, 55-58

130. Raven JA and Smith FA (1976) Nitrogen assimilation and transport in vascular land plants in relation to intracellular pH regulation. New Phytol 76, 415-431

131. Regan WS, Lambeth VN, Brown JR and Blevins DG (1968) Fertilization interrelationships in yield, nitrate and oxalic acid content of spinach. J Am Soc Hortic Sci 93, 485-492

132. Roorda van Eysinga JPNL and Meijs MQ van der (1980-1) Nitraatgehalte in glassla geoogst gedurende een etmaal. Intern verslag no 27. Proefstation voor Tuinbouw onder Glas te Naaldwijk, juni 1980

133. Roorda van Eysinga JPNL and Meijs MQ van der (1980-2) Het telen van spinazie en radijs op voedingsoplossingen met als doel het nitraatgehalte in gewas te verlagen. Intern verslag no 53. Proefstation voor Tuinbouw onder Glas te Naaldwijk, november 1980

134. Ruddell WSJ, Bone ES, Hill MJ, Blendis LM and Walters CL (1976) Gastric juice nitrate. A risk factor for cancer in the hypochlorhydric stomach. Lancet ii, 1037-1039

135. Rudert BD and Locascio SJ (1979) Differential mobility of nitrapyrin and ammonium in a sandy soil and its effect on nitrapyrin efficiency. Agron J 71, 487-490

136. Scharpf HC (1977) Der Mineralstickstoffgehalt des Bodens als Massstab für den Stickstoffdüngerbedarf. Dissertation Universität Hannover

137. Schenk M and Wehrmann J (1979) The influence of ammonia in nutrient solution on growth and metabolism of cucumber plants. Plant Soil 52, 403-414

138. Schmalfuss K and Reinicke I (1960) Uber die Wirkung gestaffelter K-gaben als KCl und K_2SO_4 auf Ertrag und Gehalt an Wasser, N-Verbindungen,K-, Cl- und S-Fraktionen von Spinatpflanzen im Gefässversuch. Z Pflanzenernaehr Dueng Bodenkd 91, 21-29

139. Schrader LE (1978) Critique-of "Factors influencing nitrate acquisition by plants: assimilation and fate of reduced nitrogen". Uptake, accumulation, assimilation and transport of nitrogen in higher plants. In: Nielsen DR and MacDonald JG, eds. Nitrogen in the environment 2, pp 101-141. New York: Academic Press

140. Schudel P, Eichenberger M, Augstburger F, Klay R and Vogtmann H (1979) Uber den Einfluss von Kompost- und NPK-Düngung auf Ertrag, Vitamin-C- und Nitratgehalt von Spinat und Schnittmangold. Schweiz landwirtsch Forsch 18, 337-349

141. Schuffelen AC, Lehr JJ and Rosanow M (1952) The technique of pot experiments. Trans Int Soc Soil Sci (Comm II and IV) 2, 263-268

142. Schuphan W (1965) Ertragsbildung und Erzeugung wertgebender Inhalts-und Schadstoffe in Abhängigkeit von der N- und P-Düngung. Landwirtsch Forsch 19 SH, 195-205

143. Schuphan W (1974) The significance of nitrates in food and potable waters. Qual

Plant Plant Foods Hum Nutr 24, 19-35

144. Schuphan W and Hentschel H (1970) Hohe Stickstoffgaben beim Spinat und ihre Folgen. Ernaehr Umsch 17, 197-200

145. Schuphan W, Bengtsson B, Bosund I and Hylmö B (1967) Nitrate accumulation in spinach. Qual Plant Plant Foods Hum Nutr 14, 317-330

146. Schütt I (1977) Zum Nitratgehalt in Spinat. Lebensm Ind 24, 318-320

147. Schwerdtfeger E (1975) Umweltbedingte Impulse als Enzyminduktoren in ihrer Auswirkung auf Pflanzeninhaltstoffe. Qual Plant Plant Foods Hum Nutr 24, 263-280

148. Siegel O and Vogt G (1974) Uber die Bildung verschiedener Stickstoffverbindungen im Spinat (Spinacia oleracea) in Abhängigkeit van Art und Menge des gebotenen Stockstoffs. Landwirtsch Forsch 27, 281-286

149. Siegel O and Vogt G (1975-1) Uber den Einfluss langsam fliessender Stickstoffquellen auf den Aufbau der Stickstoffverbindungen in Spinat und Gerste. Landwirtsch Forsch 28, 235-241

150. Siegel O and Vogt G (1975-2) Uber den Einfluss eines Nitrifikationsinhibitors auf die Stickstoffverbindungen des Spinats. Landwirtsch Forsch 28, 242-248

151. Simon C (1970) Die alimentäre Methämoglobinämie im Säuglingsalter. Ernaehr Umsch 17, 3-5

152. Singh B, Vadhwa OP, Wu MT and Salunkhe DK (1972) Effects of foliar application of S-triazines on protein, amino acids, carbohydrates and mineral composition of pea and sweet corn seeds, bush bean pods and spinach leaves. J Agric Food Chem 20, 1256-1259

153. Sissingh HA (1971) Analytical technique of the Pw method, used for the assessment of the phosphate status of arable soils in the Netherlands. Plant Soil 34, 483-486

154. Slangen JHG and Hoogendijk AW (1970) Voorschriften voor chemische analyse van gewasmonsters. Afdeling Landbouwscheikunde Landbouwhogeschool Wageningen

155. Sommer K and Mertz M (1974) Wachstum, Ertrag und Mineralstoffaufnahme von Pflanzen beeinflusst durch Ammonium oder Nitrat. Landwirtsch Forsch 27, 8-30

156. Spiegelhalder B, Eisenbrand G and Preussmann R (1976) Influence of dietary nitrate on nitrite content of human saliva: possible relevance to in vivo formation of N-nitroso compounds. Food Cosmet Toxicol 14, 545-548

157. Srivastava HS (1980) Regulation of nitrate reductase activity in higher plants. Phytochemistry 19, 725-733

158. Stephany RW and Schuller PL (1978) The intake of nitrate, nitrite and volatile N-nitrosamines and on the occurrence of volatile N-nitrosamines in human urine and veal calves. In: Walker EA, Castegnaro M, Griciute L and Lyle RE, eds. Environmental aspects of N-nitroso compounds. Lyon: International Agency for Research on Cancer (IARC Scientific Publications no 19), pp 443-460

159. Stephany RW and Schuller PL (1980) Daily dietary intakes of nitrate, nitrite and volatile N-nitrosamines in the Netherlands using the duplicate portion sampling technique. Oncology 37, 203-210

160. Tannenbaum SR (1979) Relative risk assessment of various sources of nitrite. Proc Am Meat Inst Found 1979, 67-73

161. Tannenbaum SR, Weisman M and Fett D (1976) The effect of nitrate intake on nitrite formation in human saliva. Food Cosmet Toxicol 14, 549-552

162. Tannenbaum SR, Fett D, Young VR, Land PD and Bruce WR (1978) Nitrite and nitrate are formed by endogenous synthesis in the human intestine. Science 200, 1487-1489

163. Tannenbaum SR, Moran D, Rand W, Cuello C and Correa P (1979) Gastric cancer in Colombia IV. Nitrite and other ions in gastric contents of residents from a

high-risk region. J Natl Cancer Inst 62, 9-12

164. Terman GL and Allen SE (1978) Crop yield-nitrate-N, total N, and total K relationships: leafy vegetables. Commun Soil Sci Plant Anal 9, 813-825

165. Terman GL, Noggle JC and Hunt CM (1976) Nitrate-N and total N concentration relationships in several plant species. Agron J 68, 556-560

166. Titulaer HHH (1980) Het stikstofbemestingsadvies voor spinazie. Toetsing van het voorlopig stikstofbemestingsadvies voor spinazie in de praktijk met het doel het nitraatgehalte van spinazie te verlagen. Lelystad, oktober 1980. See also: Jaarverslag 1980, Proefstation voor de Akkerbouw en de Groenteteelt in de Vollegrond te Lelystad, pp 61-63. Publikatie nr 15, september 1981

167. Tremp E (1981) Der Beitrag der einzelnen Nahrungsmittel zur täglichen Nitrataufnahme des Menschen. In: Vorträge der Informationstagung 'Nitrat in Gemüsebau und Landwirtschaft', 23. November 1981, Gottlieb Duttweiler-Institut, Rüschlikon/Zürich, pp 37-47

168. Tronicková E and Vit V (1970) The effect of fertilizers on the nitrate content in several varieties of spinach. I. Ved Pr Vyzk Ustavu Rostl Vyroby v Praze-Ruzyni 16, 119-126

169. Tronicková E and Vit V (1972) The effect of fertilizers on the nitrate content in several varieties of spinach. II. Winter spinach. Ved Pr Vyzk Ustavu Rostl Vyroby v Praze-Ruzyni 17, 273-280

170. Tychsen K (1976) Irradiance and nitrogen metabolism in spinach. Acta Agric Scand 26, 189-195

171. Ulrich A (1941) Metabolism of non-volatile organic acids in excised barley roots as related to cation-anion balance during salt accumulation. Am J Bot 28, 526-537

172. Ulrich A (1942) Metabolism of organic acids in excised barley roots as influenced by temperature, oxygen tension and salt concentration. Am J Bot 29, 220-227

173. Venter F (1978) Untersuchungen über den Nitratgehalt in Gemüse. Stickstoff 12, 31-38

174. Walker R (1975) Naturally occurring nitrate/nitrite in foods. J Sci Food Agric 26, 1735-1742

175. Wedler A (1979) Untersuchungen über Nitratgehalte in einigen ausgewählten Gemüsearten. Landwirtsch Forsch 36 SH, 128-137

176. Westvlaamse proeftuin voor industriële groenten (1978) Verslag over de proeven uitgevoerd in 1976 en 1977. Onderzoek- en Voorlichtingscentrum voor Land- en Tuinbouw, Beitem-Roeselaere

177. White JW (1975) Relative significance of dietary sources of nitrate and nitrite. J Agric Food Chem 23, 886-891
(correction in: J Agric Food Chem 25, 202 (1976))

178. Wit CT de, Dijkshoorn W and Noggle JC (1963) Ionic balance and growth of plants. Versl Landbouwk Onderz 69-15

179. Witte H (1967) Stickstoffgehalt des Spinates in der Praxis des Feldgemüsebaus. Mitt Dtsch Landwirtsch Ges 82, 1658-1660

180. Witte H (1970) Nitratgehalt des Spinates - zwei Jahre systematische Rohware-Untersuchungen. Ind Obst Gemüseverw 55, 7-11

181. World Health Organization (WHO)(1970) European standards for drinking water, 2nd Edn. Geneva, WHO

182. Wright M and Davison KL (1964) Nitrate accumulation in crops and nitrate poisoning of animals. Adv Agron 16, 197-247

183. Yem EW and Folkes BF (1954) The regulation of respiration during the assimilation of nitrogen in Torulopsis utilis. Biochem J 57, 495-508

184. Zaldivar R (1977) Nitrate fertilizers as environmental pollutants. Positive correlation between nitrates ($NaNO_3$ and KNO_3) used per unit area and stomach cancer mortality rates. Experientia 33, 264-265

185. Zaldivar R and Wetterstrand WH (1975) Further evidence of a positive correlation between exposure to nitrate fertilizers (NaNO$_3$ and KNO$_3$) and gastric cancer death rates: Nitrites and nitrosamines. Experientia 31, 1354-1355

186. Zimmermann H (1966) The influence of fertilization on the quality of spinach at various light intensities. Acta Hortic 4, 89-95